Membrane Technology in Water and Wastewater Treatment

Membrane Technology in Water and Wastewater Treatment

Edited by

P. Hillis
Northwest Water Ltd, UK

RS•C
ROYAL SOCIETY OF CHEMISTRY

EDS

SCI

The Proceedings of the Conference being held by the Water Chemistry Forum of the Royal Society of Chemistry and the European Desalination Society on Membrane Technology in Water and Wastewater Treatment at the University of Lancaster, UK, on 27–29 March 2000.

Special Publication No. 249

ISBN 0-85404-800-6

A catalogue record for this book is available from the British Library

Published by The Royal Society of Chemistry,
Thomas Graham House, Science Park, Milton Road,
Cambridge CB4 0WF, UK

For further information see our web site at www.rsc.org

Preface

These proceedings contain the papers presented at the International Conference on Membrane Technology in Water and Wastewater Treatment held at Lancaster University from the 27th–29th March 2000. The contents are set out in the order of their presentation at the conference. The meeting was organised and sponsored by the Royal Society of Chemistry Water Chemistry Forum, the European Desalination Society and the SCI Separation Science and Technology Group; with the support of the Institution of Chemical Engineers, the Chartered Institution of Water and Environmental Management and the UK Drinking Water Inspectorate.

The conference was designed to bring together membrane experts from around the world to the United Kingdom to share their experiences of the continuing development of membrane technology in the water industry. The aim of this event was to promote the use of membrane technology using MF, UF, NF and RO, for a range of water treatment applications. The conference was organised to benefit the following:

- water companies
- industrial operators
- plant designers and consultants
- R&D organisations
- membrane equipment manufacturers
- chemical suppliers
- legislative bodies

The meeting focused on all aspects of membrane technology for water and wastewater applications. It covered the treatment of ground and surface water, backwashwater, seawater and industrial and domestic wastewaters.

This book will serve as a useful reference to the status of membrane technology at the end of the second millennium and give an insight into the future of membranes as we enter the third millennium.

Peter Hillis
Editor and Conference Chairman

Contents

Fouling and Cleaning

Water Reuse

Industrial Applications

Posters

Case Studies

MEMBRANE CASE STUDIES, PAST PRESENT AND FUTURE

J. S. Taylor, Ph.D., P.E.
Alex Alexander Professor of Engineering
Civil and Environmental Engineering Department
University of Central Florida
Orlando, FL 32816
USA

S. J. Duranceau, Ph.D., P.E.
Dir. of Water Quality and Treatment
Boyle Engineering Corporation
320 East South Street
Orlando, FL 32801
USA

1 INTRODUCTION

Membrane case studies have played an essential part in the development of membrane technology for drinking water treatment. Continuing advances in regulatory constraints and aesthetic criteria for consumer water quality have driven the water community to seek new technologies, which meet these criteria. Foremost among regulatory constraints are disinfection requirements, disinfection by-product and corrosion regulations. Consumers have become aware of regulatory violation through mandated public notification, and they have always been aware of the appearance, taste and odour of drinking water. Before the requirement of advanced technologies to meet higher water quality regulations, design, construction and successful operation of conventional water plants was well established and did not require pilot studies. However, such is not the case today. Pilot or case studies in all developed countries are establishing productivity, water quality and estimated cost for advanced treatment process construction and operation.

2 OVERVIEW OF MEMBRANE PROCESSES

Understanding membrane application requires understanding of the characteristics of drinking water membrane processes. Reverse osmosis (RO), nanofiltration (NF), electro-dialysis reversal (EDR), ultrafiltration (UF) and microfiltration (MF) are the membrane processes, which have application to drinking water[11]. Combinations of membrane processes with other processes have become known as integrated membrane systems. Although a conventional NF process consists of a pre-treatment and post treatment process before and after the NF, which could be described as integrated, this is described as conventional. The coupling of a MF and a NF or coagulation, sedimentation and filtration with a NF are accepted examples of IMS's. The basic characteristics of these processes are shown in Table 1. Although many factors affect the solute separation by these process, a general understanding of drinking water application can be achieved by associating minimum size of solute rejection with membrane process and regulated contaminate.

3   REGULATIONS

The US water quality requirements determine membrane selection. Many of the regulatory constraints for drinking water can be related to control of inorganic, organic or pathogenic solutes in the finished product. The specific application of membrane processes to drinking water applications is shown in simplified format in Table 2. The word *Yes* indicates the membrane process can remove significant amounts of contaminate specified by the rule, and *No* indicates the membrane process can not remove the regulated contaminate.

Table 1 *Characteristics of Membrane Processes*

Process	Mechanism	Exclusion	Regulated Solutes		
			Pathogens	Organics	Inorganics
EDR	C	0.0001 μm	None	None	All
RO	S, D	0.0001 μm	C, B, V	DBPs, SOCs	All
NF	S, D	0.001 μm	C, B, V	DBPs, SOCs	All
UF	S	0.001 μm	C, B, V	None	None
MF	S	0.01 μm	C, B	None	None
Mechanism:	C=charge, S=sieving, D=diffusion				
Pathogens:	C=cysts, B=bacteria, V=viruses				
Organics:	DBPs=disinfection by-product precursors, SOCs=Synthetic Organic Compounds				

Examples of community water quality objectives are total hardness, taste & odour, and colour. Some community water quality objectives may be classified as secondary standards or those standards, which do not affect consumer health. Community water quality objectives are shown in simplified form in Table 3. Once the treatment objectives are known, potential membrane systems can be determined for meeting these goals. Some general statements can be made regarding current treatment concerns:

1. Diffusion controlled membranes (RO & NF) are required for control of inorganic contaminates such as total dissolved solids (TDS), total hardness (TH), chlorides, etc. and DBP precursors.
2. Charge controlled membranes (EDR) can remove TDS, TH, chlorides etc.
3. Size exclusion controlled membranes can control particles, turbidity and cysts.

Table 2 *Summary of Membrane Process Applications for Drinking Water Regulations*

US Regulation/Rule	Membrane Process				
	EDR	RO	NF	UF	MF
SWTR/ESWTR	no	yes	yes	yes	yes
CR	no	yes	yes	yes	yes
LCR	no	yes	yes	no	no
IOC	nes	yes	yes	no	no
SOC	no	yes	yes	yes	yes
Radionuclides	yes (-Rn)	yes (-Rn)	yes (-Rn)	no	no
DBPR	no	yes	yes	no	no
GWDR	no	yes	yes	yes	yes
Arsenic	yes	yes	yes	no	no
Sulphates	yes	yes	yes	no	no

SWTR -Surface Water Treatment Rule
ESWTR   -Enhanced Surface Water Treatment Rule
CR   -Coliform Rule
LCR   -Lead and Copper Rule
IOC   -Inorganic Rule (Phases I, II, IIA, V)
SOC   -Synthetic Organic Chemicals (Base Neutrals and Extractables)
DBPR -Disinfection By-Products Rule
GWDR -Groundwater Disinfection Rule

Table 3 *Summary of Membrane Process Applications for Drinking Water Regulations*

Parameter	Membrane Process				
	EDR	RO	NF	UF	MF
TDS	Yes	yes	yes	no	no
TH	Yes	yes	yes	no	no
T & O	No	yes	yes	no	no
TOC	No	yes	yes	No	no
Colour	No	yes	yes	no	no
Fe & Mn	No	yes	yes	No	no

4 REQUIREMENTS FOR CASE STUDIES

4.1 Documentation

Case studies are investigations of singular or combined processes with specific goals for production, water quality and cost. Accurate project documentation is required to develop and report pilot production and water quality. Documentation of laboratory work should be associated with a change of custody and quality control. Determination of precision and accuracy of analytical samples in the laboratory or field is essential for meaningful interpretation of results. A traceable paper trail of sample analysis should be described in project documentation and referenced in a project report. Work and sampling logs should be used for documentation of fieldwork. Work or operational logs have daily entries that describe taking operational data, mechanical repair or any other activity that is associated with operation of pilot systems. Sample logs are separate from work logs in that specific data is recorded. Typical examples of field data include pH, pressure, flow, and temperature.

4.2 Productivity

Productivity is essential to any water treatment facility. Productivity is affected by design of the membrane process and fouling. Designers can select membranes for specific treatment characteristics. Once selected, a designer can select operating conditions for that membrane process. A primary consideration affecting productivity is fouling. The four primary mechanisms of fouling are scaling, plugging, adsorption and biological growth. The primary means of controlling fouling by mechanism and unit operation are shown in Table 4. General Comments can be made regarding pre-treatment requirements.

1. Scaling control is typically required for all RO/NF membrane systems in either surface or groundwaters and is achieved by acid and/or antiscalent addition.
2. Plugging control is typically required for all RO/NF membrane systems in either surface or groundwaters and is achieved by feed water turbidities and SDI's less than 0.2 NTU and 2 respectively.
3. Bio-fouling control is typically required for aerobic surface or groundwaters and is achieved by NH_2Cl or addition of other bactericidal agents.
4. Organic fouling can occur in surface water systems with TOC > 3-6 mg/L and is typically reduced by coagulation, sedimentation and filtration. However, the significance of organic fouling is not known.

Table 4 *Fouling Control by Pre-treatment System*

Mechanism	Process				
	A/AS	MF/UF	CSF	NH_4Cl	AOC Removal
Scaling	+	-	-	-	-
Plugging	-	+	+	-	-
Adsorption	-	+	+	-	-
Bio-fouling	-	-	-	+	+

4.3 Evaluation of Membrane Systems

Performance of membrane units can be measured in the field or laboratory by functioning membrane systems in the form of (a) small cells, (b) bench units or (c) pilot plants. Small cells have historically been used to test film characteristics and have not yet been shown to be representative of actual production. A testing protocol for small cells has been published[1]. There results indicated that the Phase II MCLs for DBPs could be met in four of the five waters tested, the rate of productivity decline was exceptionally high and that DOC was diffusion controlled. While the DBP results were comparable with previous studies, the productivity results were not. Their results have shown that small cells are suited for membrane screening quality studies but are not adequate for productivity assessment.

Membrane manufacturers use 4"x40" elements, 2.5"x40" elements or 2.5"x20" elements in pilot studies for plant scale up. The preliminary results of the information collection rule (ICR) were presented at the 1999 WQTC in Tampa, FL[2]. In excess of forty membrane studies have been done through the United States to comply with the ICR requirements. The studies typically consist of flat sheet laboratory work; single element field work, multi-staged pilot plants and full scale plant documentation. The studies have shown that DBP precursor removal requirements were exceeded for haloacetic acids for either stage 1 or 2 MCLs, but that the ninety-fifth percentile of THMs exceeded the stage 2 MCL (40 ug/L). The higher THMs were attributed to high bromide concentrations (>30 ug/L) and variable bromide rejection (70 % to 0 %) by membranes.

Membrane case studies require that mass transport of water and solutes through the membranes be described quantitatively. The equations used to describe flow through a single element are shown in (1) through (7) with reference to the membrane element shown in Figure 1.

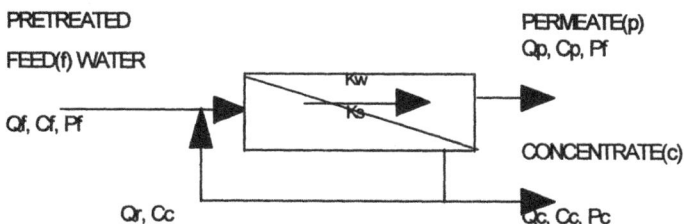

Figure 1 *Basic Diagram of Mass Transport in a Membrane*

Mass transport in pressure driven membrane processes can be described as convection or diffusion controlled. Models for describing mass transfer in membrane systems have been presented by several investigators.[11,4,9] These equations can be used to predict permeate water quality of any membrane system. Although membrane systems have been shown to produce water quality that exceeds most regulatory requirements, models as shown in equation (8) or (9) are essential for predicting the cost and performance of membrane systems from pilot plant data. Equation (8) is used to describe diffusion controlled mass transfer, which includes inorganics, i.e., alkalinity, hardness, TDS, sodium, chlorides, etc. Equation (9) is used to describe sieving controlled mass transfer, which includes TOC, DBP precursors, most SOCs and organics in general.

However, TOC has been shown as diffusion controlled in a surface water application at Tampa FL for CA and CTF membranes[10]. The TOC from the CTF membrane was so low that no limit for flux and recovery was projected for precursor control, however flux and recovery were limited for the CA membranes as TOC rejection was significantly less.

Existing research has clearly shown consistent production and concentrate disposal are the limiting constraints for membrane systems. Concentrate disposal is essentially a regulatory problem but consistent production is a research and development problem. There is simply inadequate information on the fouling of membranes by moderate organic surface waters.

Reference Equations:

$$F_W = K_W(\Delta P - \Delta \Pi) = \frac{Q_p}{A} \qquad (1)$$

$$\left[\frac{C_s - C_p}{C_f - C_p} \right] = e^{\frac{F_w}{k}} \qquad (6)$$

$$F_S = K_S \, \Delta C = \frac{Q_p C_p}{A} \qquad (2)$$

$$r = \frac{Q_r}{Q_f} \qquad (7)$$

$$R = \frac{Q_p}{Q_f} \qquad (3)$$

$$C_p = \frac{K_S C_f e^{\frac{f_w}{k}}}{F_w \left(\frac{(1-r)(2-2R)}{2+2r-R} \right) + K_S e^{\frac{f_w}{k}}} \qquad (8)$$

$$Q_f = Q_c + Q_p \qquad (4)$$

$$Q_f C_f = Q_c C_c + Q_p C_p \qquad (5) \qquad C_p = \Phi C_v \qquad (9)$$

ΔC = Concentration gradient $(M/L^3), ((C_f + C_c)/2 - C_p)$

C_f = Feed stream solute concentration (M/L^3)

C_c = Concentrate stream solute concentration (M/L^3)

C_p = Permeate stream solute concentration (M/L^3)

C_s = Solute concentration at the membrane surface (M/L^3)

k = Diffusion coefficient from the surface to the bulk (L^3/L^2t)

ΔP = Pressure gradient (L), $((P_f + P_c)/2 - P_p)$

$\Delta \pi$ = Osmotic pressure (L) $((\pi_f + \pi_c)/2 - \Delta_p)$

K_W = Solvent mass transfer coefficient (L^2t/M)

F_W = Water flux (L^3/L^2t)

F_S = Solute flux (M/L^2t)

K_S = Solute mass transfer coefficient (L/t)

Q_f = Feed stream flow (L^3/t)

Q_c = Concentrate stream flow (L^3/t)

Q_p = Permeate stream flow (L^3/t)

R = Recovery

A = Membrane area (L^2)

r = Recycle ratio

ϕ = Sieving pass coefficient

Fouling indices can be used to indirectly estimate pre-treatment requirements for membrane systems[6]. These investigations have shown that fouling indices do not statistically correlate to the rate of productivity decline in diffusion controlled membrane systems. Regardless, fouling indices can be used to get a crude estimate of membrane pre-treatment requirements. The silt density index (SDI), modified fouling index (MFI) and the mini-plugging factor index (MPFI) are shown in equations (10), (11) and (12).

$$SDI = \frac{100\left(1 - \frac{t_i}{t_f}\right)}{T} \qquad (10)$$

$$MFI = (Q V)^{-1} \qquad (11)$$

$$MPFI = \frac{Q}{t} \qquad (12)$$

Where: t_i=*time to collect initial 500 ml;* t_f = *time to collect final 500 ml;* T = *time between* t_i *and* t_f *(15 min);* Q=*flow;* V=*volume*

These indices that can be used to predict the required pre-treatment for RO or NF membrane processes. These indices are determined by monitoring volume collected verses time of filtration through a 0.45 μm filter under a constant pressure of 30 psig. Approximate values of RO and NF indices are shown in Table 5.

Table 5 *Site Raw Water Quality*

Parameter	Units	TWD	BG	MELB	ESL	FM
UV-254	(cm-1)	0.8895	0.209	N/A	0.2	0.5
Colour	(cpu)	188	56	480	20	200
TOC	(mg/L)	19.3	6.6	47	4	20
SDS-TOX	(μg/L)	922	310	9,000	NA	NA
SDS-TTHM	(μg/L)	1,426	253	1,866	200	>1000
SDS-HAA6	(μg/L)	2,032	234	3,561	100	>800
Alkalinity	(mg/L as $CaCO_3$)	67	148	110	120	100
Total Hardness	(mg/L as $CaCO_3$)	90	215	150	120	150
TSS	(mg/L)	4.7	41	N/A	10	5
TDS	(mg/L)	151	332	430	300	400
SPC	(cfu/100 ml)	1,816	ND	ND	ND	ND
MIB	(ng/L)	<2.0	ND	ND	ND	ND
Geosmin	(ng/L)	8.4	ND	ND	ND	ND
Iron	(mg/L)	0.35	0.0317	0.82	0.3	0.3
Bromide	(mg/L)	0.063	0.07	ND	<0.1	<0.1
Manganese	(mg/L)	0.029	<0.002	N/A	ND	ND
Chloride	(mg/L)	11.6	33.5	100	80	40
Sulphate	(mg/L)	11.8	60.5	2	80	20
Silica	(mg/L)	10.4	6.3	ND	15	15
Nitrate	(mg/L)	0.16	15.3	N/A	<5	<1
Pesticides	(g/L)	<1	4.7	N/A	N/A	N/A

These indices have been shown to not statistically correlate to flux or mass transfer decline during membrane operation but are used as guides for membrane pre-treatment. Collection of indice data is useful to estimate pre-treatment, can be easily and inexpensively done, and could provide data that relates pre-treatment requirements of varying water qualities.

Consistent productivity is essential to successful membrane applications. Productivity is can be measured by determining K_w during time of operation and the determining the rate of decline of productivity from the associated slope as shown in equation (14).

Productivity can be measured by monitoring flux over time if the pressure gradient is held constant. Monitoring K_w over time normalises flux based on pressure and provides a truer assessment of productivity. A plot of K_w vs. time curve from pilot plant data can be used to predict the cleaning frequency for membrane plants using equation (14) or (15). Both approaches have been shown satisfactory[8]. Such a curve is shown in Figure 2. This data was generated from a pilot plant study in Florida and was used to predict the cleaning frequency based on 10 % productivity or K_w decline.

$$K_W = \frac{F_W}{\Delta P - \Delta \Pi} \qquad (14)$$

$$K_W = K_{W_0} - m \bullet t$$

$$\frac{1}{K_w} = \mu \left[\frac{\Delta P}{F_w} + \mathrm{E}t \right] = \frac{1}{K_{wo}} + A_2 t \qquad (15)$$

The intercept of the curve shown in Figure 2 is the initial MTC and the run time before cleaning can be easily predicted by using equation (14) as shown in the following example. The slope of the curve in Figure 2 is the rate of MTC decline during production. Such data and calculations can easily be generated from any pilot plant studies and with required water quality measurement could easily be used to meet the requirements of the ICR, as the example below provides:

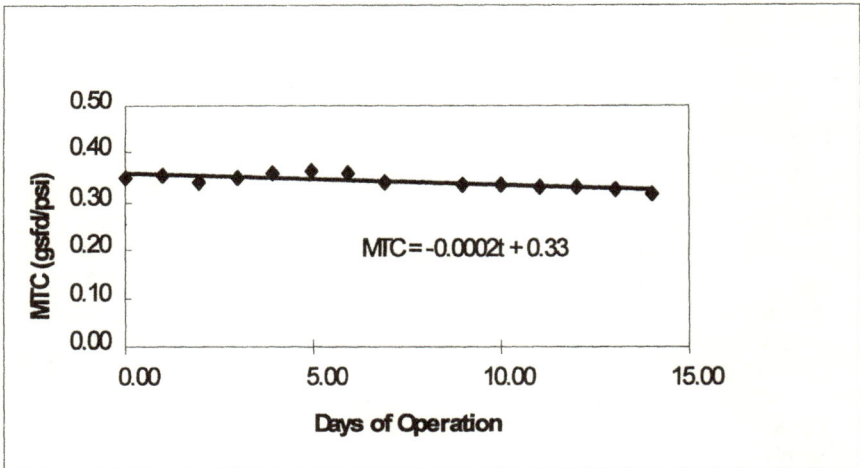

Figure 2. *MTC VS Time of Operation*

$$K_{wo} = 0.3317 \; \frac{gsfd}{psi}$$

$$10\% \; decline = (0.10)K_{wo} = 0.03317 \; \frac{gsfd}{psi}$$

$$rate \; of \; K_w \; decline = 0.0002 \; \frac{gsfd}{psi-d}$$

$$Cleaning \; Frequency = \frac{0.03317 \; \frac{gsfd}{psi}}{0.0002 \; \frac{gsfd}{psi-d}} = 158 \; days$$

5 CASE STUDY SELECTION

The selected sites for presentation were associated with research projects[12,13,3]. All of the case studies could be described as integrated membrane systems (IMS's), however the data from each of the membranes used in the IMS's can be utilized for application of singular membrane units receiving similar water quality. The site raw water quality is shown in Table 6.

Treatment objectives for each site were established by reviewing the site raw water quality with respect to the regulatory constraints as shown in Table 7. All sites were classified as surface waters. All RO/NF membranes require control of plugging and precipitation or scaling. There is no acceptable method of determining the effects of organic or biological fouling other than conducting pilot studies, which assess each fouling mechanism separately. The case studies by site are shown in Table 8. Whereas the complete removal of any one mechanism of fouling is highly unlikely, the reduction of a particular mechanism of fouling is highly likely. The comparison of productivity declines for various scenarios for fouling control can provide an assessment of separate fouling mechanisms.

Table 6 *Fouling Index Approximations for RO/NF*

Fouling Index	Range	Application
MFI	$0 - 2 \; s/L^2$	Reverse Osmosis
	$0 - 10 s/l^2$	Nanofiltration
MPFI	$0 - 3 \times 10^{-5} \; L/s^2$	Reverse Osmosis
	$0 - 1.5 \times 10^{-4} \; L/s^2$	Nanofiltration
SDI	$0 - 2$	Reverse Osmosis
	$0 - 3$	Nanofiltration

Table 7 *Treatment Objectives by Site*

Site	Treatment Objectives							
	Cysts	DBP	TOC	Nitrate	SOCs	Hardness	Colour	Turbidity
BG	X	x	x	x	x	x	x	x
TWD	X	x	x				x	x
MELB	x	x	x				x	x
ESL	x	x	x				x	x
FM	x	x	x			x	x	x

Table 8 *Pre-treatment and Nanofilter by Site*

Site	Pre-treatment							Nanofilter				
	A/AS	CSF	CS	ILC	GF	UF	MF	NH₄Cl	BAC	TFC1	TFC2	CA
BG-1	x	X								x		
BG-2	x					x				x		
BG-3	x						x			x		
BG-4	x			x		x				x		
BG-5	x		x			x				x		
TWD-1	x	X								x		
TWD-2	x	X						x		x		
TWD-3	x	X										x
TWD-4	x	X						x				x
TWD-5	x						x			x		
TWD-6	x						x	x		x		
TWD-7	x						x					x
TWD-8	x						x	x				x
TWD-9	x			x			x			x		
TWD-10	x			x			x	x		x		
TWD-11	x			x			x	x				x
ESL-1	x	X						x		x		
ESL-2	x	X						x			x	
ESL-3	x	X						x				x
MELB-1	x	X								x		
MELB-2	x	X									x	
MELB-3	x						x			x		
MELB-4	x								x	x		
MELB-5	x								x		x	
FM	x				x					x		

6   RESULTS AND DISCUSSION

6.1 Bowling Green, Ohio

The cleaning frequencies for the BG-membrane system are shown in Table 9. CSF pre-treatment resulted in projected NF cleaning frequencies varying from four days to infinity. Cleaning frequency increased with flux and recovery. No cleaning was project at 7 gsfd and 65 % recovery. MF pre-treatment resulted in projected NF cleaning frequencies varying from 19 days to infinity and was superior to CSF or UF pre-treatment. UF fouling prohibited UF pre-treatment from being realised. ILC-UF resulted in NF cleaning frequencies varying from 27 days to infinity and UF cleaning frequencies of 8 to infinity. Both ILC and CS using $FeCl_3$ resulted in reduced UF fouling. Final selection of a BG-membrane system would require cost evaluation of capital and operation cost of NF. BG Cleaning frequencies varied with pre-treatment system and NF flux and recovery. Typically, coagulation and reduced flux and recovery eliminated or increased NF cleaning frequency.

The primary fouling mechanisms for NF at BG is indicated as organic fouling or possibly biological fouling. Although CSF, CS or ILC pre-treatment does remove active biota by enmeshment in sweep floc, it is unlikely that these processes as used at BG would remove enough substrate to limit biological growth on the membrane. BG finished water quality is shown in Table 10. All BG water quality goals were met or exceeded by each membrane system. All water quality exceeds existing plant water quality. The water quality data indicates that pre-treatment systems did not greatly affect finished water quality. This would be expected for relatively low concentrations of diffusion controlled or size exclusion solutes.

Table 9 *Cleaning frequency for the different processes (days)*

	MF		UF		NF		
	Max	Min	Max	Min	Flux (gsfd)/recovery(%)		
					7/65	7/85	14/65
CSF-NF	x	x	x	x	∞	28	9
MF-NF	8 [1]	36 [2]	x	x	∞	271	19
UF-NF	x	x	1		x	x	x
CS-UF-NF	x	x	∞		∞	x	x
ILC-UF-NF	x	x	8 [3]	∞[4]	x	27	∞

1) Flux =40 gsfd, BW frequency=40 min; 2) Flux =20 gsfd, BW frequency=20 min; 3) low and high dose of coagulant
4) dose of 21 ppm of FeCL3, flux=40gsfd

Table 10 *Finished BG Plant and Water Quality*

		BG-Plant	CSF-TFC	MF-TFC	ILC-UF-TFC
TOC	mg/L	4.5	0.55	0.26	0.3
THMSDS	ug/L	111	2	2	3
HAASDS	ug/L	86	12	3	2
Turbidity	NTU	0.22	0.05	0.05	0.05
Nitrate	mg/L	21.1	3.38	1.66	0.99
Pesticides	ug/L	4	N/S	<0.4	N/S

6.2 City of Tampa, Florida

Eleven different membrane systems were investigated at TWD. The raw water source is a surface water that is highly organic and feasibly required control of scaling, plugging, organic adsorption and biological fouling. Hence, the eleven membrane systems at Tampa were intended to evaluate the control of all mechanisms.

The fouling evaluations for TWD membrane systems MFs applications are presented in Tables 11, 12 and 13. The fouling of the Memcor MF with and without in-line coagulation (ILC) is shown in Table 11. Memcor cleaning frequency increased as flux and backwash frequency increased. Flux is indicated to be more important to MF fouling than backwash frequency. ILC significantly reduced MF fouling and indicates that organic fouling of the polypropylene membrane was the controlling mechanism. Ferric salt significantly fouled the MF. Both alum and PACl were effective for reducing MF fouling. Extrication of alum solids from the MF was difficult. The TWD Memcor MF data indicates that organic fouling and membrane surface chemistry significantly affected MF fouling.

The Zenon cleaning frequencies for the TWD are presented in Table 12 for varying MF flux and recovery. Cleaning frequency decreased as flux and recovery increased. Estimated Zenon cleaning frequencies varied from 3 to 86 days.

Cleaning frequencies for the Zenon MF receiving ferric sulphate coagulated solids are shown in Table 13. Cleaning frequency decreased with increasing coagulant dose, which indicates an adverse reaction between the iron solids and the membrane surface. The least cleaning resulted for no ferric sulphate addition. The Zenon data indicates that the surface characteristics of the membrane were the most significant factor affecting MF fouling at TWD. Biological fouling of the Zenon MF was not considered to be significant given the low run times, however the Zenon MF is used extensively in wastewater systems without any control of biofouling. The Zenon MF system has no pressure vessel, which enhances solids handling capability and the difficulty of bacterial inactivation on the membrane surface.

TWD NF cleaning frequencies are shown in Table 14 by membrane, NH_4Cl addition and NF flux and recovery. The CA NFs were destroyed by biological degradation, which was controlled by NH_2Cl and acid addition. The CA NFs were replace and operated successfully. These results indicate that the CA NF fouled much less than the TFC NF. Flux and recovery showed no effect on CA NF fouling, whereas flux and recovery did effect the TFC membrane. Typically, TFC fouling increased as flux and recovery increased. NH_4Cl addition significantly reduced NF fouling of TFC NFs and was necessary to avoid biological degradation of CA NF. Consequently the data indicates that biofouling is significant at TWD and that organic fouling is not indicated as a significant fouling mechanism at TWD.

The TWD plant and membrane system water quality is shown in Table 15. All membrane systems water quality exceeds plant water quality and meets all TWD water quality goals. The TOC, DBP and turbidity water quality is not surprising. The CA NF rejection of organics was not as great as TFC rejection of organics. Geosmin and MIB are removed to less than the level of detection by the TFC NF, and reduced to 3 and 1 ng/L by the CA NF. This is another indication of the improved water quality produced by membrane technology.

Table 11 *TWD MMF and C/MMF Conditions and Projected Cleaning Frequencies.*

Flux	BW Frequency	Coagulant	Dose	Cleaning frequency
gsfd	Min		meq/L	days
14	15	x	x	36
16	15	x	x	21
17	20	x	x	29
20	20	x	x	14
20	40	x	x	7
40	20	x	x	4
In-line coagulation				
17	20	PACl	0.15	130
20	20		0.15	
15	20	Alum	0.60	190
15	20		0.66	50
15	40		0.22	12
15	40		0.66	4
20	40	Ferric	0.40	1
20	40		0.20	3

Table 12 *ZMF Cleaning Frequency for Varying Flux and Recovery*

	Cleaning frequency (days)				
		Flux (gsfd)			
		20	30	32	40
	75	56	23	18	10
Recovery	85	40			
%	90	86	12		
	95		3		

Table 13 *C/ZMF Operational Conditions and Projected Cleaning Frequencies.*

Test	Flux	Recovery	BP Frequency	BP Duration	Air Flow	Coagulant dose	Cleaning frequency
	gfd	%	min	sec	cfm	meq	Days
1	26	80	12	20	17	0.22	12
2	26	95	12	20	17	0.22	12
3	26	80	12	20	17	0.67	12
4	32	80	12	20	17	0.97	7
5	29	80	12	20	17	1.79	2
6	29	80	12	20	17	1.79	2
7	29	80	12	20	17	3.00	2

Table 14 *Nanofilter Productivity Summary*

Membrane System			Cleaning Frequency (days)			
PTMT	NF	NH$_2$Cl	7/65	7/85	14/65-75	14/85
CSF	CALP		>270	>270	>270	>270
MMF	CALP	X	>270	>270	>270	>270
ZMF	CALP	X	>270	>270	>270	>270
C/MMF	CALP	X	>270	On-going	-	-
C/ZMF	CALP	X	>270	>270	>270	>270
CSF	LFC1	X	No fouling -		-	No fouling
	LFC1		No fouling -		-	20
MMF	LFC1	X	-	-	-	5^1
	LFC1		8	4	10	2
ZMF	LFC1	X	No fouling 21		No fouling 5	
	LFC1		No fouling -		-	3
C/MMF	LFC1	X	-	-	-	On-going

^{1} – *Additional experiment scheduled. Monochloramine dose was low and pre-existing levels of biogrowth are thought to have reduced effectiveness of this experiment*

Table 15 *TWD Plant and Membrane Systems Water Quality*

	TOC	THM	HAA	Turbidity	Geosmin	MIB	TON	TDS
	mg/L	ug/L	ug/L	NTU	ng/L	ng/L		mg/L
Plant CSF	3.2	90	50	0.14	N/A	N/A	10	271
CSF-CALP	0.5	57	19	0.08	N/A	N/A	6	90
ZMF-CALP	0.8	60	57	0.09	3.2	< 1.0	2.9	82
C/ZMF-CALP	0.6	35	23	0.1	1.3	4.9	1.3	205
MMF-CALP	0.5	34	23	0.09	1.6	9.2	2.4	203
CSF-LFC1	< 0.5	3.3	4	0.09	< 1.0	< 1.0	1.5	26
ZMF-LFC1	< 0.5	6	7	0.09	< 1.0	< 1.0	2.4	35
MMF-LFC1	< 0.5	4	27	0.08	< 2.0	< 1.0	2.2	12

6.3 East St. Louis, Illinois

Six membrane systems have been investigated at East St. Louis Illinois. Productivity results for the ESL systems consist of CSF water fed into three separate NFs. Two TFC NFs and one CA NFs have been investigated to date. Membrane systems consisting of CA UF membranes are being investigated at ESL but CA UF and NF membranes have been destroyed due to oxidation. No evidence of chemical oxidation has been observed. Both the NF and the UF systems employ addition of chlorine during operation, however biological oxidation is possible if material is not removed from the CA membrane surface. Excessive silt was observed on the CA NF, autopsy of the CA UF membrane is not complete.

The TFC membranes are shown to foul at a greater rate than the CA membranes. Scaling significantly reduced productivity. Reducing the feed pH to 5, which may have caused aluminium precipitation in the NFs, controlled calcium carbonate scaling. Only CSF pre-treatment results have been reported today. As shown in Table 16, these results indicate that NF cleaning frequency decreases as NF flux and recovery increase, and that CA NFs foul less than TFC NFs does. These results also indicate that CA membranes are susceptible to degradation and not as durable as TFC membranes.

Table 16 *ESL Cleaning frequency by System and NF Flux and Recovery*

	NF Cleaning Frequency (days)			
System	Flux (gsfd)/recovery(%)			
	10/55	10/75	15/55	15/75[1]
CSF-TFC1	30	10	8	No fouling
CSF-TFC2	40	10	7	No fouling
CSF-CA	TBD	20	12	70

[1] Feed pH reduced to 5.0

The ESL water quality is shown in Table 17. The ESL project is currently in progress however the data indicates that the CA membranes do not reject TOC as well as the TFC membranes. CA NFs may produce waters that are more likely to exceed stage II DBP MCLs than TFC NFs may.

Pathogen rejection data was collected from the ESL investigation and used to create the data shown in Table 18. As each ESL pilot unit was challenged separately by cysts, bacteria and phage, measured rejection of the organisms was possible. This data was coupled assuming 2.5, 1.5 and 0.5 log rejection of cysts, bacteria and phage by CSF and compared to pathogen rejection by membrane system and CSF. This data indicates pathogen rejection by membranes is superior to conventional CSF.

Table 17 *Finished ESL Plant and Water Quality*

		ESL-Plant	CSF-TFC1	CSF-TFC2	CSF-CA
TOC	mg/L	5	0.3	0.2	0.8
THMSDS	ug/L	30[*]	22	31	48
HAASDS	ug/L	20[*]	20	35	50
Turbidity	NTU	0.13	0.05	0.06	0.05

[*] Monochloramine addition controlled DBP formation

Table 18 *Calculated Pathogen Log Rejection for ESL*

Parameter	CSF NF-1	CSF NF-2	CSF NF-3	UF NF-1	UF NF-2	ILC UF NF-3
Cysts	7.0[*]	4.1[*]	6.9[*]	TBD	TBD	TBD
Bacteria	5.5[*]	2.6[*]	5.9[*]	7.8	4.9	7.7
Virus	4.8[*]	2.3[*]	4.7[*]	7.1	4.6	7.0
NPDOC	1.0	0.7	1.1	1.0	0.7	1.1
Inorganics	1	0.3	0.3	1	0.3	0.3

[*] Assumes 2.5, 1.5 and 0.5 log removal for Cysts, Bacteria and Viruses respectively by CSF

6.4 City of Melbourne, Florida

The Melbourne membrane systems consisted of two different TFCs NFs, which were preceded by parallel MF, CSF and GAC pre-treatment. The conventional NFs systems consisting of A/AS and static MF and NF could not operate for more than 2 hours due to plugging. Comparison of cleaning frequencies indicates that the alum CSF pre-treatment produced the least fouling water or a 20-day runtime. BAC pre-treatment was utilized to remove substrate and control biological fouling. The short BAC pre-treatment run was due to high turbidity and plugging of the NF membranes. However this data indicates that BAC did not effectively control biofouling and additional filtration following BAC would be necessary to control NF fouling.

Assuming MF pre-treatment reduced plugging, an assumption can be made that the increased runtime following CSF was due to a reduction in organic or biofouling. CSF does partially reduce bacteria loading onto the NF, but alum CSF water is highly biologically active. Although an exact determination of fouling mechanism is not possible, it can be stated that higher TOCs from groundwaters given to conventional NFs application run for six months to one year before cleaning. Such sources are anaerobic and have a reduced and different bacterial population. Additionally in the limited data reported in these studies, there has been no instance where addition of a bactericidal agent did not increase runtime.

Table 19 *Melbourne Chemical Cleaning Frequencies*

System	K_w Decline Rate	Time to 15% Decline
	(gfd/psi/day)	(days)
GACB-E	-0.00248	12
MF-E	-0.00352	9
ACSF-E	-0.00150	20
GACB-F	-0.01283	2

Assumes initial Kw of 0.2 gfd/psi

Table 20 *Melbourne Plant and Water Quality Summary*

Parameter		Plant	Permeate				
			GAC-E	MF-E	ACSF-E	GAC-F	ACSF-F
DOC	mg/L	10.5	0.2	0.5	0.5	<0.2	0.2
THMFP	ug/L	360	18	31	45	20	20
HAA(5)FP	ug/L	262	5	18	18	10	11
Colour	cpu	9	< 1	1	1	< 1	<1
Turbidity	NTU	0.3	0.1	0.1	0.1	0.2	0.1
HPC	cfu/ml	2428	173	1129	146	1123	545
TDS	mg/L	460	162	230	163	149	95
CaH	mg/L $CaCO_3$	143	45	45	18	26	15

6.5 City of Fort Myers, Florida

Fort Myers is a full-scale plant, which was evaluated as part of an AWWARF project[3]. The membrane system consists of pumping water from the Caloosahatchee River to rapid infiltration basins (RIBs) for infiltration into a shallow aquifer. Water is pumped from the aquifer to a conventional NF plant, which employs acid and anti-scalent addition to control scaling. The production results for the Fort Myers system are reported from 1993-1997.

The utilisation of the available membranes for Fort Myers is shown in Figure 3. The plant was initially constructed for 12-MGD capacity and begins operation in 1993 at slightly over 60 % of the available capacity. Capacity in this sense is defined as a percentage of utilized membranes that are available membranes for operation. The data in Figure 3 shows the utilisation of membranes increased from slightly over sixty percent to slightly over eighty percent from 1993 to 1997.

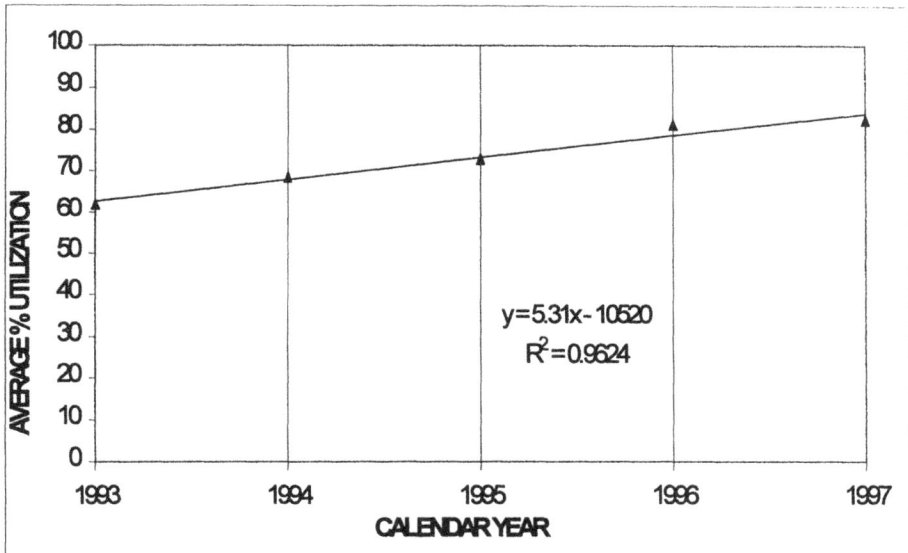

Figure within chart:
$$y = 5.31x - 10520$$
$$R^2 = 0.9624$$

Figure 3 *Fort Myers Membrane Utilization Over Time*

The data in Figure 4 shows the average production from 1993 to 1997. This data shows that the average production decreased from slightly over 7 MGD to slightly more than six MGD from 1993 to 1997. Consequently utilisation increased and production increased from 1993 to 1997. Increasing utilisation of membranes (membrane surface area) and decreasing production is an indication of membrane fouling in an operating facility.

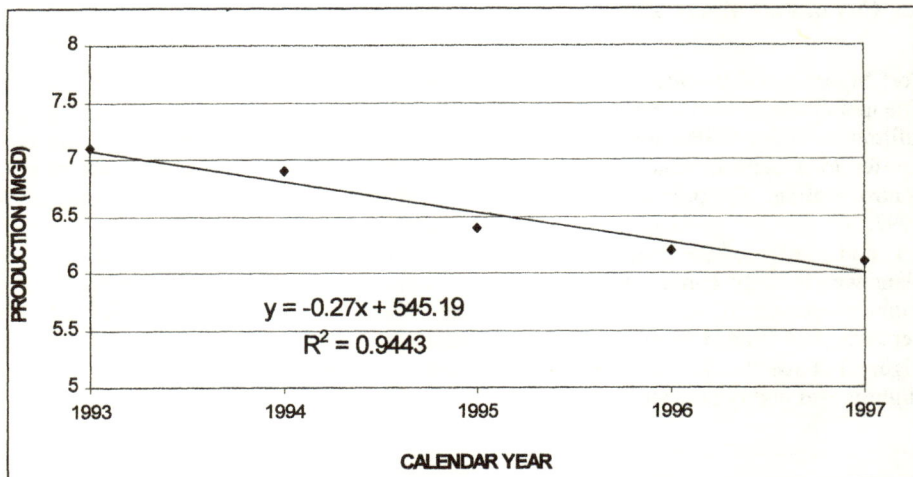

Figure 4 *Fort Myers Production Over Time*

The Fort Myers systems recovery is shown by stage over time of operation in Figure 5. These results indicate that first stage recovery decreased from approximately seventy percent to fifty percent over five years of operation. Second stage recovery increased from twenty percent to 30 percent and third stage recovery remained constant over the same period. The results indicate colloidal deposition and removal in the first stage, which caused extensive fouling of the first stage. The increased resistance in the first stage increased second stage productivity. This observation is similar to first and second stage productivity relationships when a flow restriction is places on a stage one permeate line, which is commonly done to prevent excess stage one productivity.

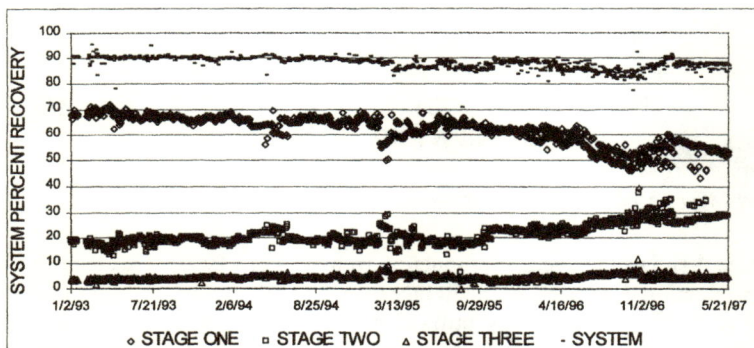

Figure 5 *Fort Myers Recovery by Stage Over Time*

The Fort Myers normalised flux and MTC is shown in Figures 6 and 7. Both of these figures clearly show a consistent fouling over time of operation. The mechanism of fouling is plugging. The water quality data in Table 6 for the Fort Myers NF shows iron and sulphides are present in a slightly aerobic source. The Fort Myers wells are less than 30 feet deep and subject to intermittent aerobic conditions. Such an environment is consistent with the chemical precipitation of iron or sulphur, and biological growth. Over five years, there were 161 estimated cleanings of individual banks in the Fort Myers NFs. The rapid fouling and chemical conditions of the Fort Myers feed water indicates plugging as the primary mechanism of fouling.

The Fort Myers raw, feed and permeate water quality is shown in Table 21. Although the Fort Myers productivity was unsatisfactory, the water quality was good. Only THMs were observed to be higher than the Stage II MCL. The NPDOC while below 1 mg/L was high for typical NF permeate. Both NPDOC and THMs may decrease when the productivity problem is solved.

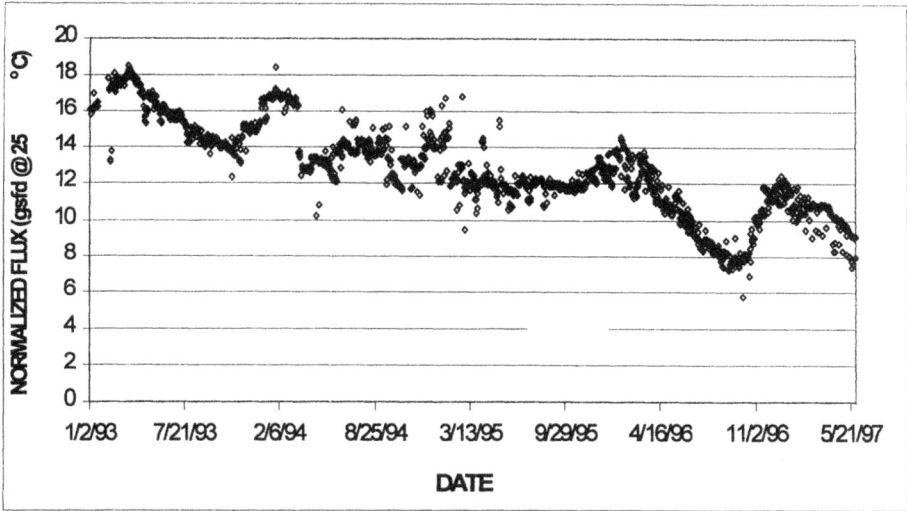

Figure 6 *Fort Myers Normalised Flux Over Time*

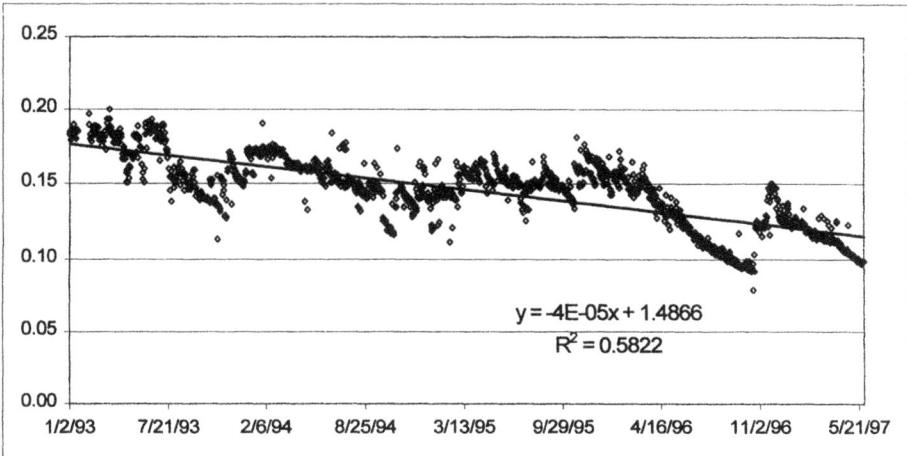

Figure 7 *Fort Myers MTC Over Time*

Table 21 *Fort Myers Water Quality*

Parameter		Raw	Feed	Permeate
NPDOC	mg/L	22	18	0.8
Colour	CPU	173	85	5
Ca	mg/L	53	75	18
Mg	mg/L	24	14	3
H_2S	mg/L	0	2	0
DO	mg/L	8	0.14	0.14
Fe	mg/L	0.3	0.3	<0.2
Alkalinity	mg/L $CaCO_3$	145	208	42
SO_4	mg/L	21.3	20.1	15.9
Bromide	mg/L	0.334	0.458	NR
Na	mg/L	27	35	31
Cl	mg/L	42	60	33
Turbidity	NTU	0.68	0.16	0.03
THM	ug/L	1095	735	69
HAA	ug/L	801	377	18

7 CASE STUDY COSTS

Costs for case studies would include planning costs, labour costs, permits, equipment rental and/or purchase, electricity, chemicals and laboratory analyses. Not unlike other engineering planning, design and construction costs, there are contingencies that arise and ancillary expenditures that will be incurred during the course of project implementation. The costs expended for the case studies described herein ranged between $250,000 USD and $400,000 USD, and included several funding sources (utility support, government research funds, university matching funds, in-kind donated funds). These costs should be viewed as value engineering and investment costs, as the information derived from performing detailed pilot investigations will impact bottom-line operation and maintenance costs. Also, pilot investigations can also be used to implement training of personnel and other resources prior to construction and operation of the full-scale facility.

For example, a pilot investigation may find that cleaning frequencies are twice of that predicted, resulting in design of additional pre-treatment facilities (prior to construction) and increased operation budgets at the onsite of start-up. If a detailed pilot investigation had not been conducted, then significantly higher costs (upwards of 20 to 30 percent of the original design estimate) would have to be expended to change-order and or perform a new design and construction step after acceptance of the original construction occurred. In addition, the operation costs would have been projected to be much lower than actual; causing a significant alteration of the total amortised cost for which bonding is typically based upon. The key point here is to say that pilot investigations are mandatory for confirming costs and performance of membrane processes for a site-specific application.

8 FINDINGS AND CONCLUSIONS

1. RO, NF, EDR, UF and MF are the five membrane processes, which have major drinking water application. Case studies are required to determine design criteria for applications of membranes in drinking water treatment. Membranes are necessary to meet the regulatory constraints of increased water quality criteria; consequently, membranes offer a broad range of regulatory compliance to utilities.
2. Membranes can be effectively used singularly or in combinations of pre-treatment unit operations and conventional NF/RO systems to enhance NF/RO productivity. Correlation of membrane process and contamination removal can be made by pore and contaminate size.
3. Plugging, scaling, organic adsorption and biological growth are major fouling mechanisms for membranes, which must be assessed in applications. Site specific pilot investigation is required to determine significance of fouling mechanisms. Basic fouling mechanisms can be assessed by incremental evaluation of combinations of fouling mechanisms. RO/NF fouling can be assessed by observation of MTC decline over time of operation; plugging and scaling must be controlled in every RO/NF application.
4. RO/NF fouling in groundwater systems is typically significantly less than in surface water systems although TOC in CSF surface waters is less than conventional RO/NF TOC in groundwater systems. This is consistent with biological fouling of aerobic sources and decreased significance of organic fouling.
5. Organic fouling is a significant fouling mechanism for RO/NF applications but is typically not as significant as biological fouling.
6. Biological growth is a very significant fouling mechanism in every RO/NF application involving aerobic raw water sources (surface waters and shallow wells), and must be directly assessed for RO/NF fouling. Biological and chemical oxidation must be controlled for CA membranes in any treatment application.
7. NF pathogen rejection at ESL by TFC NF exceeded conventional CSF pathogen rejection. Combinations of membranes (Miss) exceed pathogen rejection by conventional treatment.

9 SUMMARY

The implementation of membrane case studies is important to support decision making activities when designing, constructing and initiating a membrane process for potable water production. The value of the study is often dependent on the manner in which the study was performed. Site specific variables impact the results of the pilot investigation. Pilot investigations assist in confirming design criteria, water quality goals, concentrate disposal parameters, membrane replacement estimates, operation costs, cleaning costs, maintenance estimates and construction costs. Pilot investigations are also used to familiarise and train operations personnel in the understanding of the technology relative to full-scale operations prior to start-up. Pilot investigations also can provide information related to concentrate water quality that can assist in determining disposal or reuse options. Fouling is perhaps the most important variable that must be delineated in a pilot investigation, as many have found that without site specific pilot investigations, the construction of those facilities will often result in membrane plants that do not produce design water quantities, typically because fouling issues were not identified prior to construction and operation of the facility.

References

1. Allgeier, S.C., Summers, R. S. 1995. *"Evaluating NF For DBP Control With The RBSMT"*, Journal AWWA. 87:3:87, (March 1995).

2. Allgeier, S.C. 1999. *"Analysis of Nanofiltration Under the ICR."* Proc. AWWA Water Quality and Technology Conference, Tampa, FL, Nov. 1999.

3. AWWARF. 1998. *"Investigation of Integrated Membrane Systems"*, Quarterly Reports, 1996-1998, work in progress, Denver, CO

4. Duranceau, S.J. 1990. *"Modeling of Mass Transfer and Synthetic Organic Compound Removal in a Membrane Softening Process."* Doctoral Dissertation, University of Central Florida, Orlando, FL.

5. Lovins W. A.., Taylor J. S.,Kozik R., Abbasedegan, M., LeChaevallier M. and Aty, K. 1999. *"Multi-contaminant Removal by Integrated Membrane Systems"*, Proceedings of AWWA Water Quality and Technology Conference, Nov 1999.

6. Morris, K. M., 1990. *"Predicting Fouling in Membrane Separation Processes,"* Master Thesis, University of Central Florida, Orlando, Florida

7. Duranceau, S.J. 1990. *"Modeling of Mass Transfer and Synthetic Organic Compound Removal in a Membrane Softening Process."* Doctoral Dissertation, University of Central Florida, Orlando, FL.

8. Mulford, L. A., Taylor J. S., Nickerson, D. M. and Chen Shaio-Shing, *"NF performance at full and pilot scale."* Journal AWWA, Vol. 91, No. 6, June 1999.

9.Sung, Larry. 1993 *"Modeling Mass Transfer in Nanofiltration,"* Doctoral Dissertation, University of Central Florida, Orlando, FL.

10. Reiss, C. R., Taylor J. S., Owen C., and Robert C. 1999. *"Diffusion-Controlled Organic Solute Mass Transport in Nanofiltration Systems."* Proceedings of AWWA Water Quality Technology Conference, Tampa, FL, Nov. 1999.

11. Taylor, J. S.; Duranceau, S.J.; Barrett, W.M.; Goigel, J.F. 1989. *"Assessment of Potable Water Membrane Application and Research Needs,"* AWWA Research Foundation Report, Denver, CO.

12. USEPA. 1992. *"Reduction Of Disinfection By-Product Precursors By Nanofiltration"*, USEPA/600/SR-92/023, April 1992.

13. USEPA. 1998 *"Removal of Multi-Contaminates by Integrated Membrane Systems"*, Quarterly Reports for Project in Progress, 1996-1998, WERL, DRWD, Cincinnati, OH

SWRO – THE LARGEST PLANT IN BRITISH WATERS

Mr N Marsh and Mr J Howard

Jersey New Waterworks
Mulcaster House
Westmount Road
St Helier
Jersey
Channel Islands

Ms F Finlayson and Dr S Rybar

Weir Westgarth Ltd
149 Newlands Road
Cathcart
Glasgow
G44 4EX

1 ABSTRACT

The paper describes the operational experience with the newly constructed SWRO plant in Jersey. The plant replaced a multi-stage flash distillation plant, commissioned in 1970 by Weir Westgarth, which was operational until 1996. At the time of supply the MSF plant was considered to be the most energy efficient desalination process with a Gain Ratio of 12 to 1 to minimise running costs.

Projected Operating Costs for the new SWRO plant are compared with the actual running costs as well as with actual running costs for the original MSF Plant. The environmental impact of both installations is also addressed.

For this particular site the benefits of membrane technology are clearly demonstrated.

Keywords: *Seawater; Reverse osmosis; Operating costs; Environmental impact*

2 INTRODUCTION

Jersey is the most southerly of the Channel Islands, measuring approximately 14.5 km by 8 km and is situated approximately 24 km from France. The geology of the island is varied and complex but includes ancient shales with intrusions of granite, andesites, and rhyolites, overlaid by sand deposits.

The annual average rainfall is 847.4 mm (133 mm year average).

The Jersey New Waterworks Company Limited (JNWW) was founded in 1882 and is the oldest registered company operating in the Island. The resident population of the island is approximately 85,000 which rises to about 120,000 in the summer. The JNWW currently supplies approximately 85% of the island.

3   HISTORY

3.1   Operation, Water Supply and Demand

Natural aquifers account for only 3% of the raw water supply. Unfortunately, some of the highest yielding catchments have the smallest reservoirs. A network of raw water pumping stations has been developed to fill the reservoirs from remote stream sources and also provide a flexible means to supply Handois and Augres Water Treatment Works.

The maximum storage available is 2,677,340 m^3. This represents only 4 months demand. The maximum daily demand up to June 1998, was 27,534 m^3 (6.06 mg/d), in 1994. This maximum daily demand indicates a shortfall of 5,223m^3/day and based on this, the capacity of the SWRO plant was fixed at 6,000 m^3/day.

3.2   How the Plant Performed

The Multi-Stage Flash Distillation Plant (MSF) was constructed between 1968 and 1969 by Weir Westgarth and was commissioned in 1970. The rated output of the MSF plant was 6,715 cubic metres (1,500,000 imperial gallons) per day. The plant has produced a total of 6,550,000 m^3 of product water. The plant was used extensively during the summer droughts of 1976, 1989 and 1990, where 950,000 m^3, 877,000 m^3 and 1018,000 m^3 of water were produced respectively.

3.3   Process Description

The existing distiller was a high performance unit (design GOR 12 to 1), to minimise energy consumption. It consisted of 47 stages, 44 stages were designed to recover heat and 3 stages to reject heat. Each stage operated at progressively increasing negative pressure, until Stage 47 was almost at full vacuum. At Stage 47 the water boiled at 37 °C. On entering each stage a portion of the brine flashed and via demister pad and condensed of tube bundles. The resulting steam passed upwards to condense on tube bundles to produce the distillate. The feed water was dosed with sulphuric acid, to reduce calcium carbonate deposits, anti scalant to assist in keeping solids in suspension, and a de-foaming agent.

The distillate being of high purity was chemically aggressive and so lime was added to modify the hardness and thus prevent damage to the pumping main to Val de La Mare Reservoir. The distillate was then blended with natural water sources prior to being pumped to the water treatment works.

The plant was decommissioned in 1998, 28 years after commissioning, as part of the contract for the new Reverse Osmosis plant.

3.4   Economics of the MSF Plant

The capital cost of the plant, including 5,500 m of distillate main was £1.2M in 1970.
The Weir Westgarth Contract for the MSF distiller, boilers, oil tanks, and all mechanical plant and instrumentation was £775,000.

The fuel cost to run the plant excluding all other costs at 1998 prices would be £4,083 per day.

4   CHOICE OF NEW TECHNOLOGY

Advances in membrane technology over recent years have had a significant affect on the choice of the process for the proposed plant. The plant is a single purpose plant with a low load factor, with Reverse Osmosis (RO) providing the most attractive process option in terms of flexibility of operation, space requirements and minimal visual impact in a tourist area.

The RO process offers the following advantages, compared to thermal processes:
-    lower energy requirement
-    shorter start-up and shut down times
-    lower maintenance requirement
-    operational flexibility
-    no atmospheric pollution

5   NEW PLANT DESCRIPTION

The new sea water desalination system consists of the following main stages (Figure 1):
-    sea water screens and intake pumps
-    quarry pool pumps
-    pre-treatment with auxiliaries
-    high pressure system pumps and RO trains with energy recovery
-    product pumps and product water stabilisation

The plant is designed to have the daily output of 6000 m^3/day of permeate with a quality of less than 400 mg/l TDS, based on a feed water quality of 37,734 mg/l TDS. Full output can be maintained over the sea water temperature range of 7.5°C to 19.5°C. The sea water salinity is in reality slightly higher and TDS based on individual ion analysis are around 38,000 mg/l. Feed water conductivity fluctuates around 54.0 mS/cm.

5.1   Sea Water Intake Screens and Intake Pumps

The sea water intake station is located on the shoreline. Water from the sea passes through the submersed tunnel to the intake pit. Due to the high amount of seaweed, present during rough seas, a macerator is installed in the intake pit. Three vertical intake pumps - two in duty one in standby - pump seawater to the quarry pool tank via an old cement coated cast iron pipeline. An, automatic, self cleaning filter is installed before the seawater enters the quarry pool tank with a filtration rate of 800 microns. The filter, which has operated satisfactorily since April, is fully automatic and self cleans on a pre-set differential pressure or timer. The quarry pool has an estimated capacity of 32,000 m^3 and in the past debris in the seawater settled in the bottom of the quarry pool. The role of strainer is to eliminate further sedimentation and decay of particles larger than 800 microns.

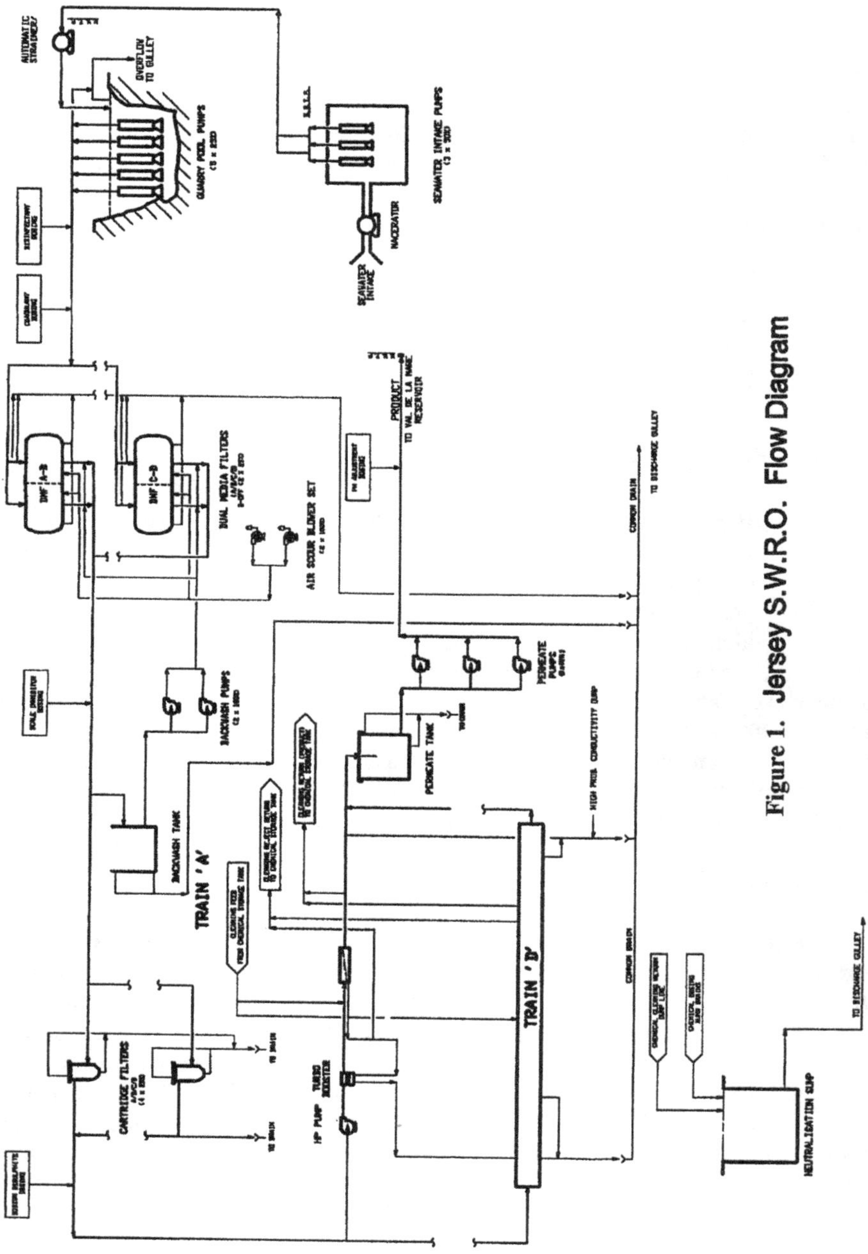

Figure 1. Jersey S.W.R.O. Flow Diagram

5.2 Quarry pool pumps

Under normal continuous flow conditions the quarry pool level will fluctuate by approximately one meter. The quarry pool pumps are designed to supply the plant with sufficient flow and pressure up to 7.6 meters below the normal operating level. This allows drainage of the quarry pool after long term plant shut down. Five quarry pool pumps are provided, four duty and one stand by.

5.3 Pre-treatment

The role of the pre-treatment is to filter seawater to a quality acceptable to feed into RO membranes and includes:
> DMF`s; backwash pumps and backwash tank; air scour blowers; cartridge filters; pre-treatment chemical dosing.
> The dosing of flocculant upstream of the media filters agglomerates the suspended solids and colloidal matter.

Under automatic control the flocculant is regulated to guarantee quick and optimum flocculation. The flocculant used is an organic type, based on the coagulation tests carried out on site in November last year and Weir Westgarth's operational experience from existing installation's (Gibraltar)[1]. In addition a set of compatibility tests was carried out to find if there is a potential danger of increased membrane fouling due to the reaction between organic coagulant and membrane. Test and operational results indicated no increased fouling potential due to the use of organic flocculant.

System sterilisation is based on intermittent shock chlorination using sodium hypochlorite. It was expected that shock disinfecting - chlorinating would be required once every two weeks for about an hour. During the first three months of operation this has had to be carried out weekly and for 3 - 4 hours.

Flocculated colloidal matter and suspended solids are removed in dual media filters. Four such filter chambers are provided in two pressure vessels. The filters normally operate at a filtration velocity of 8.8 m/h. With one chamber in backwash, the velocity in the remaining chambers reaches 11.8 m/h.

The projected operation time between two backwash cycles is 24 hours, corresponding to a dual media filter layer capacity for suspended solids of approximately 2.9 kg/m^2 or 2.8 kg/m^3 of filtration media at 13 mg/l of suspended solids in feed water.

Operational data gathered to date have indicated an average quarry pool SDI of 28 and an average DMF outlet quality of 4.5 - 4.8 SDI.

The filtered water enters a common header connected to four (33% duty) 10-micron cartridge filters. These provide the final pre-treatment step. The cartridge filter outlet is connected to the high pressure pumps suction header.
Scale inhibitor is added upstream of the cartridge filters by proportionally controlled dosing pump sets (3 x 50% pumps).

5.4 RO Membranes

4 x 1,500 m^3/day membrane streams are provided. Each stream consists of a high pressure pump; turbo-booster and set of membranes to produce 1,500 m^3/day of permeate. The high pressure pump selected is of horizontal centrifugal design.

The turbo-booster utilises the high pressure brine rejected from the reverse osmosis membranes. Transferring pressure energy in the brine reject stream to the feed stream, reducing the electrical power required during normal operation.

Plant conversion of 45% was selected to ensure the power requirement was within the 2MVA power supply. With 45% conversion the projected specific energy requirement of the plant is 5.75 kWH/m^3 for process load alone (pre-treatment and RO part). It is important to point that all pumps used in this installation are designed and manufactured by WEIR PUMPS Ltd.

Normalising plant performance against the first day of operation using ASTM D4516-85 (Reapproved 1989) and a temperature correction factor from the membrane manufacturer assess train performance.

Membrane performance has been as good as predicted by the manufacturer with normalised salt passage of 0.27 - 0.3% at design flow rates.

5.5   Control System

The system provided offers the flexibility and performance normally attributed to a distributed control system (DCS), and is designed to provide minimum operator intervention and maintain the necessary interlocks for safe operation.

To facilitate the control of the new SWRO plant a PLC based system has been selected with a PC based topology incorporating a SCADA software package for ease of operation.

The system architecture has three major components:
1.   Programmable Logic Controller (PLC)

A host PLC with direct connections to field signals and communications interfaces to PC based visualisation system (VS) or main machine interfaces (MMI) and remote terminal units (RTU).
2.   Remote Terminal Units (RTU)

As the geographic layout of the plant is diversified field cabling is minimised by introducing the use of remote terminal units (RTU) which are located in field junction boxes to remotely collect the field instrumentation signals and relay them back to the PLC via a bus system.
3.   Visualisation System (VS) or Man Machine Interface (MMI)

Visualisation system (VS) or man machine interface (MMI) run on PC's with communication to the PLC. The PC runs a software package and represents the process configuration as MIMIC diagrams with individual displays of plant items or equipment and faceplates for control points. A common point database resides on the system and is shared between the VS/MMI and the PLC, however the VS/MMI is contained and configured separately on the PC. The VS/MMI also offers trending, facilities, alarm handling, operator prompts and messages and data storage.

A direct link between Weir Westgarth Ltd and the plant control room enables process supervision or operational data transfer at any time

5.6   Energy Efficiency and the Environment

For comparison we have used a price of 5p per kWh of electricity. This gave a designed daily energy running cost of approximately £2,025 per day (compared to £4,083 for the MSF plant) (Table 1).

The new RO plant avoids the burning of 45,400 litres of heavy grade oil per day and consequent emission of 2.2 tonnes of SO_2. This provides a significant environmental improvement. Chemical costs were in the region of 8p/day for the MSF plant. The RO plant has similar chemical costs but utilises less hazardous compounds.

The new plant generates no gaseous or odorous emissions. The main effluent streams are brine and cleaning fluids. The characteristics of liquid effluents as well as the rejected flows can be summarised as follows:

-        brine is the major liquid effluent produced and its concentration in chemical compounds is approximately double the concentration of the feed seawater. The chemistry of the brine varies according to the chemistry of the raw seawater. The wave motion rendering it harmless quickly disperses it

-        organic coagulant is used in the process as an alternative to ferric chloride to avoid pollution of the sea with staining iron salts. The organic coagulant chosen is biologically degradable.

-        sodium hypochlorite used for shock disinfecting is neutralised before discharge to the sea

-        preservation chemicals and cleaning chemicals are neutralised in the neutralisation pit before they are discharged to the sea

6   SUMMARY

The MSF Distillation Plant constructed between 1968 and 1969 by Weir Westgarth Ltd was decommissioned and replaced by a new desalination plant with capacity of 6000 m^3/day, also designed and built by Weir Westgarth Ltd. The contract was signed 5th of January 1998, with the plant handed over to the client on 29th June 1999. The design, construction, commissioning and reliability tests were executed within 18 months. The plant employs the latest water treatment technologies and minimises operating costs.

Table 1 *Running costs*

ROCESS	OUTPUT (M³/DAY)	ENERGY COST (£/DAY)	PROCESS&CLEANING CHEMICAL	CART. FILTER REPLACEMENT	TOTAL COST
MSF	6517	4083 (Fuel Oil) 0.626/m³	£0.08/m³	-	£0.706/m³
SWRO design	6000	2025 (Electricity) 0.338/m³	£0.08/m³	£0.01/m³	£0.428 m³
SWRO actual	6000	2022(Electricity) 0.337/m³	£0.08/m³	£0.006/m³	£0.423 m³

References

1   EG Darton AG Turner EDA Canagua 1996
2   W R Querns (Private Communications)

DRINKING WATER SOURCES IN KUWAIT

M. Safar and Y. Al-Wazzan

Water Desalination Department, Water Resources Division,
Kuwait Institute for Scientific Research
PO Box 24885,
Safat 13109,
Kuwait

ABSTRACT

Kuwait's water resources are limited to desalinated water, brackish water and wastewater. In Kuwait, seawater desalination is the main source of freshwater required by all demand sectors.

Other marginal supplies come from groundwater and bottled water. Freshwater is commercially produced by seawater desalination by two main processes, i.e., multistage distillation flash (MSF) and reverse osmosis (RO).

This paper contains detailed information regarding drinking water in Kuwait in terms of system configuration, production rate and water quality standards. It also focuses on the implementation of membrane separation systems (i.e., reverse osmosis) for the production of freshwater from both seawater and brackish water.

Keywords: Multistage flash, membrane, bottled water, groundwater.

1 INTRODUCTION

The history of civilisation is closely linked to the availability of water, which has always been the primary source of economic growth, being directly related to food production and thus of the communities' concern. Invariably, the history of nations reveals that the survival of any civilisation is directly related to water management. The shortage of potable water, especially in hot climatic zones, causes social discord and the disruption of economic development[1].

The World Bank has pointed out in it water management strategy that the water situation in the Middle East is precarious. Per capita renewable water resources have fallen from approximately 3500 m^3 to less than 700 m^3 [2] Thus; water must now be viewed as a limited resource not as a sectored input. The shortage in natural water supply for domestic purposes is more acute for the Arabian Gulf area, where the demand for water increases annually at a rate of 3 or more percent[3]

Natural resources of freshwater in Kuwait are very limited because Kuwait is situated in an arid coastal region characterised by high temperature, decreased humidity, little rainfall and high evaporation rate. Therefore, Kuwait has always had to search for other sources to secure freshwater to meet with its growing demands. Kuwait in the past relied mainly on rainwater found near the surface in shallow wells, but due to the growth of population, that scant source became no longer sufficient to cater to the growing demand. So, in 1939, Kuwait turned to the Shaat Al-Arab for freshwater supply brought by boats, and a primitive storage and distribution network was established. The water transported at a rate of 8500 IGPD increased to 80000 IGPD by the end of 1946[4]. This situation prevailed until the influx of oil wealth and the first oil shipment in 1946. Kuwait then had the funds necessary to invest in modern water production facilities that could cater to freshwater demand.

This paper describes in detail all sources of freshwater in Kuwait in terms of system configuration, production rate and water quality standards. It also focuses on the implementation of membrane separation systems (i.e., reverse osmosis) for the production of freshwater from both seawater and brackish water.

2 DRINKING WATER SOURCES IN KUWAIT

Freshwater in Kuwait can be classified into four main categories:
underground water, bottled water, desalinated water by reverse osmosis (RO), and desalinated water by multistage flash (MSF).

2.1 Underground Water

Underground water in Kuwait can be classified into two main categories: brackish and fresh underground waters. The brackish water exists in the Kuwait Group Aquifer and Dammam Aquifer stretching east of the Arabian Peninsula and slightly sloping towards the Arabian Gulf. The main location of the brackish water wells are Sulaibia fields, Shagaya fields, Um-Qudair field and Al-Wafra and Al-Abdalia fields currently utilised by the Kuwait Oil Company, in addition to wells in the agricultural areas of Al-Wafra and Al-Abdalia (Figure 1). In 1997, the quantity of brackish water withdrawn, including that used for blending with desalinated water, was 65.7 MIGPD[4].

As for fresh underground water, the first well that had a relatively large freshwater capacity was discovered in Hawally in 1905. Limited quantities were discovered during 1962 in both Al-Rawdatain and Um-Al-Aish fields. Pumping operations for these fields was commissioned in 1962 and their estimated natural reserve was about 40000 MIG. But after the Iraqi invasion, the Um-Al-Aish field was destroyed and, as a result, its' production was stopped. As for the Al-Rawdatain field, the usual rate of production capacity is about 1 MIGPD, which, when necessary, could be raised to 2.5 MIGPD for a period of 10 to 15 days at a maximum of three times a year to preserve the quality of water.

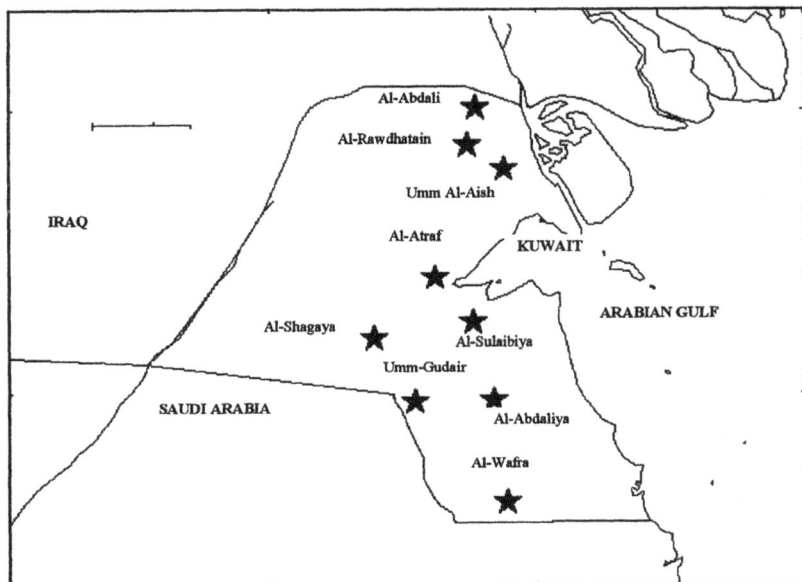

Figure 1. *Location map of the underground water in Kuwait*

2.2 Bottled Water

Bottled water is a generic term that describes all water sold in containers. Natural mineral water is by definition and regulation untreated product water extracted from a naturally protected source. It differs fundamentally from bottled spring water and tap water, which rely upon treatment as a means of ensuring portability[5]. The main difference between bottled water and ordinary drinking water is that the bottled water has certain characteristic contents and concentration of certain minerals. This is obtained from natural underground sources either by natural flow or drilled bore holes and bottled with or without some treatment process (i.e., filtration and sterilisation). The mineral content of bottled water depends solely on the rocks and the duration that the water comes into contact with the geological surroundings.

Temperatures in desert climates reach about 50°C or higher during summer and causes countries like Kuwait to consume large quantities of natural mineral water. In Kuwait, there are more than 70 brand names of natural mineral water imported from all over the world and sold in the local markets. However, there is one brand name produced locally by Al-Rawdatain Natural Mineral Water Bottling Company.

Al-Rawdatain Natural Mineral Water Bottling Company was setup in November 1980 on a 50 square kilometre mineral water field, 100 kilometres north of Kuwait City with an operational capacity of 9.68 MIGPY. The actual yearly production of Al-Rawdatain water in the beginning of 1983 was about 4.8 MIG. Production of Al-Rawdatain natural mineral water is increasing every year, reaching a value of 7.26 MIG in 1995[6].

Table 1 presents the chemical analysis for the bottled water available in Kuwait's market. The data presented in this table were taken from labels found on the water bottles. These labels usually show ten elements that represent the main key parameters,

Table 1 Water Analysis of Bottled Water Available in the Kuwaiti Market

No	Name	Country	pH	Bicarbonates	Chlorides	Sulphates	Nitrates	Fluoride	Calcium	Magnesium	Potassium	Sodium	TDS
1	A	Bahrain	7.00	16.00	40.00	0.01			1.40	1.50		28.06	90.00
2	AA	Bahrain	7.10	28.00	41.00	17.60	3.96	0.68	3.20	0.48	1.00	58.00	118.00
3	AB	Bahrain	7.10	28.00	41.00	17.60	3.96	0.68	3.20	0.48	1.00	58.00	118.00
4	AC	Bahrain	7.10	28.00	41.00	17.60	3.96	0.68	3.20	0.48	1.00	58.00	118.00
5	AD	Bahrain	7.20	40.00	57.00	8.00	5.00	0.70	2.40	1.50		55.00	144.00
6	AE	Belgium	7.70	204.00	4.00	18.00	3.50		67.60	2.00	0.20	1.90	201.00
7	AF	Canada	6.70		1.00	4.70			10.00	0.85	0.60	1.20	49.00
8	AG	Egypt		107.36	28.00	16.50		0.60	7.04	8.25	17.00	34.00	200.00
9	AH	Eygpt		325.00	50.00	50.00			60.00	26.00	5.00	62.00	440.00
10	AI	Eygpt		122.00	36.00	18.00		0.45	8.00	9.60	18.00	42.00	200.00
11	AJ	France	7.00	65.30	8.40	6.90	6.30		9.90	6.10	5.70	9.40	
12	AK	France		256.00	23.00	6.00	16.00		86.00	3.00	1.70		316.00
13	AL	France		357.00	4.50	10.00	3.80		78.00	24.00	1.00	5.00	
14	AM	France		263.00	12.00	13.00	0.60		80.00	8.00	2.00	8.00	370.00
15	AN	France		258.00		105.00			91.00	19.90		7.30	
16	AO	India	7.00	28.00	16.00	24.00			16.00	14.00		38.00	60.00
17	AP	Iran	7.10	195.00	5.00	16.00	0.00	0.40	55.00	15.00	13.60	1.40	
18	AQ	Iran	8.00	188.00	5.00	10.00	0.00	0.03	56.00	12.96	1.40	4.25	
19	AR	Italy	6.40	29.80	3.20	5.90	3.30	0.05	6.90	1.20	0.90	3.40	49.60
20	AS	Italy	8.03	183.00	0.80	16.50	3.50		36.00	18.50	0.44	0.65	
21	AT	K.S.A	7.00	45.00	10.00	25.00	5.70	0.70	10.00	4.90	0.30	10.00	110.00
22	AU	K.S.A	7.00	41.00	15.00	22.00	8.30	0.90	9.52	5.06	0.60	19.00	101.00
23	AV	K.S.A	7.00	45.00	10.00	25.00	4.00	0.70	10.00	1.20	0.30	10.00	110.00
24	AW	K.S.A	7.10	36.70	28.10	11.80	1.80	0.60	12.80	2.30	1.40	18.40	
25	AX	K.S.A	7.10	30.00	28.00	50.00	5.00	0.68	24.00	5.00	1.00	19.00	140.00
26	AY	K.S.A	7.20	57.00	30.00	27.00	15.00	0.70	20.00	5.00	3.00	20.00	170.00
27	AZ	K.S.A	7.20	105.25	62.90	10.80	2.87	0.70	2.40	1.22	1.32	80.75	220.00
28	B	K.S.A	7.20	30.00	35.00	8.00	4.00	0.75	2.50	1.00	2.00	38.00	125.00
29	BA	K.S.A	7.20	110.00	28.80	11.50	0.00	0.70	13.50	3.50	1.80	24.60	177.00
30	BB	K.S.A	7.20	45.00	15.00	38.00	6.00	0.80	18.00	5.00	1.50	28.00	180.00
31	BC	K.S.A	7.30	66.00	32.00	28.00		0.72	10.00	1.20	0.36	50.00	150.00
32	BD	K.S.A	7.30	44.00	29.00	27.00	3.52	0.65	8.80	3.40	1.60	30.80	165.00
33	BE	K.S.A	7.40	55.00	8.00	20.00		0.70	18.00	5.00	2.00	0.00	103.00
34	BF	K.S.A	7.40	35.00	26.00	36.00	4.00		10.00	3.00	0.50	32.30	140.00
35	BG	K.S.A	7.40	77.80	23.90	35.30	0.00	0.70	22.50	7.90	9.40	17.30	180.00
36	BH	K.S.A	7.50	18.20	34.80	19.30	13.30	0.75	7.60	1.20	0.54	31.20	160.00
37	BI	K.S.A	7.50	25.00	50.00	70.00	8.00	0.75	36.00	8.00	1.50	30.00	190.00
38	BJ	K.S.A	7.90	46.00	50.00	90.00	9.00	0.68	44.00	8.00	1.80	30.00	265.00
39	BK	K.S.A.	7.00	20.00	10.00	35.00	2.00	0.70	10.00	4.00	0.30	10.00	110.00
40	BL	K.S.A.	7.15	30.50	20.50	19.00	5.00	0.70	12.60	4.50	1.10	13.00	109.00
41	BM	K.S.A.	7.20	90.50	52.25	12.50	3.23	0.70	10.60	3.70	0.50	30.50	110.00
42	BN	Kuwait	7.80	155.00	9.00	22.00	8.00		45.00	7.00	4.00	9.00	
43	BO	Kuwait	7.80	155.00	9.00	22.00		0.80	45.00	7.00	4.00	9.00	
44	BP	Kuwait	7.80	155.00	9.00	22.00	8.00		45.00	7.00	4.00	9.00	
45	BQ	Lebaeon	7.20	180.00	10.00	40.00		0.25	46.00	16.00	3.00	20.00	300.00
46	BR	Lebanon	7.30	81.00	9.00	4.00	4.00	0.60	21.00	9.00	1.00	6.00	135.00
47	BS	Lebanon	7.90	105.20	5.10	10.90	1.80	0.01	31.30	5.20	0.50	3.50	130.00
48	BT	Oman	7.80	210.00	30.00	42.00	8.00		52.00	20.00	1.60	21.00	C≈480
49	BU	Oman	7.80	210.00	30.00	42.00	8.00		52.00	20.00	1.60	21.00	C≈480
50	BV	Oman	7.90	220.00	28.00	44.00	5.00		55.00	21.00	2.00	19.00	C≈470
51	BW	Oman	7.90	220.00	28.00	44.00	5.00		55.00	21.00	2.00	19.00	C≈470
52	BX	Scotland	7.80	163.00	7.50	6.00	1.00	0.10	35.00	8.50	1.00	6.00	136.00
53	BY	Spain		191.00	39.70		4.00		25.60	23.60		30.50	
54	BZ	Swiss		306.00	11.50	211.00	1.80	1.40	115.00	40.00	1.80	19.90	718.00
55	C	Tunisia	7.00	232.00	26.00	24.00	13.00		70.00	11.00	3.00	12.00	306.00
56	D	Tunisia		255.00	50.00	31.00	23.00		88.00	7.00	4.00	29.00	380.00
57	E	Turkey	6.60	22.00			4.40		3.60	1.40			
58	F	Turkey	8.00	8.20	5.40	8.00	1.30		23.20	5.40	0.30	3.70	90.00
59	G	U.A.E	8.05	87.00	70.00	11.00	4.40		6.00	28.00	2.00	31.00	
60	H	U.A.E.	7.40		36.00					5.00	1.00	20.00	
61	I	U.A.E.	7.50	34.00	36.00	4.00	0.20		9.00	4.00	1.00	19.00	
62	J	U.A.E.	7.60	70.00	43.00		1.40	0.20	4.00	11.00	4.00	29.00	
63	K	U.A.E.	8.05	87.00	70.00	11.00	4.40		6.00	28.00	2.00	31.00	
64	L	U.A.E.	8.10	85.40	51.00	38.00	3.50		4.40	30.60	2.00	20.25	
65	M	U.A.E.	8.10	155.00	56.00	44.00	5.90		19.20	29.00	1.30	21.00	
66	N	U.A.E.	8.25	80.60					5.00	23.50	1.70	12.50	
67	O	U.A.E.	8.25	80.60					5.00	23.50	1.70	12.50	
68	P	U.A.E.	8.25	80.60					5.00	23.50	1.70	12.50	
69	Q	U.A.E.	8.25	80.60					5.00	23.50	1.70	12.50	
70	R	U.A.E.	8.25	80.60					5.00	23.50	1.70	12.50	
71	S	U.S.A.	7.53	0.00					5.90	3.10	2.10	0.00	83.00
Average values			7.47	109.63	26.75	27.60	5.11	0.61	26.87	10.48	2.48	21.59	180.62

and they are calcium, magnesium, potassium, sodium, sulphate, nitrate, chloride, fluoride, hydrogen ion concentration (pH), total dissolved solid (TDS) and the bacteriological quality of the bottled water. These labels also show the production and expiration dates and the origin.

These criteria should be available on all bottled water so that customers can make the right decision on what brand and what quality is best for them. However, this is not usually the case. Some labels placed on the bottled waters presented in Table 1 show only five elements (Nos. 60 and 66 to 71), whereas others do not show the TDS (Nos. 11, 13, 15,17,18, 20, 24, 42 to 44, 53, 57 and 59 to 70).

Out of the 71 different water brands (Table 1), there were 45 brand names imported to Kuwait from the Gulf States, which can be attributed to the close distance between these countries and Kuwait. It was also noticed that the bottled water produced in the United Arab Emirates (UAE) has high pH values (around 8.25), whereas the bottled water produced by the other Gulf states are found to be below 8.0.

Table 2 shows that there are many bottled waters that have identical chemical analysis but are marketed under different trade names.

Table 2. *Bottled Waters with Identical Chemical Analysis*

Serial No. in Table 1	No. with Identical Chemical Analysis
2- 4	3
48, 49	2
50, 51	2
59, 63	2
66-70	5

2.3 Water Desalination by Reverse Osmosis

2.3.1   Doha Experimental Sea Water Reverse Osmosis Project. In 1979, a cooperation agreement was signed between the State of Kuwait and the Federal Republic of Germany. According to the agreement, both parties constructed an experimental water plant at the Doha site with a capacity of 3000 cubic meters per day using the reverse osmosis (RO) method. The plant contained three systems, which differ in design, membrane configuration and chemical treatment. The German party continued participation till the end of 1987 while the Kuwaiti party continued the research program. The most important results of that program were the reliability of RO in seawater desalination under the local conditions of Kuwait. The RO units at the Doha site were completely destroyed during the Iraqi invasion of Kuwait in 1990. New two single-stage seawater RO units with a total capacity of 600 m^3/d were installed.

Each RO unit is designed to operate independently to produce 300 m^3/d using raw seawater from a common beachwell system. One unit utilises a twin hollow fibre membrane, whereas the other unit utilises a spiral wound, thin film membrane. The seawater feed is withdrawn from two beachwells drilled at 30m depth near the beach at Doha. It consists of the two well pumps, a complete dosing system, two independent RO trains and a complete post-treatment system. The seawater feed can be chlorinated by dosing with Cl_2 gas. De-chlorinating of the seawater upstream of the membranes should be ensured. An acid dosing system was also installed to adjust the pH value to 7.0 to ensure a negative Stiff & Davis Index value. Furthermore, an antiscalant is dosed to inhibit carbonated and non-carbonated scaling on the membrane. The pre-treated water then passes a fine filter system consisting of three-bag fine filters with a mesh size of 5 μm.

A multistage, centrifugal, horizontal, high-pressure pump pressurises the pre-treated seawater to the required operating pressure. The pressurised water is guided to the membrane stacks, which are equipped with suitable membranes to separate the salt in the brine stream from the freshwater permeate in one stage. The number of membranes are selected to enable the rated output of 300 m^3/d per train (spiral wound train consists of five pressure vessels each housing six elements, and hollow fibre train consist of eight pressure vessels each housing a twin membrane element). The permeate of each train flows to the holding surge/suck back tank located above the module racks and from there by gravity to the flushing tank. Each train is equipped with a Pelton wheel turbine to recover the energy from the reject brine, which is discharged from the RO membrane.

2.3.2 Brackish Water Desalination by Reverse Osmosis (RO) Brackish groundwater exists in reasonable quantities, but it is barely renewable. During 1997, the daily production of brackish water in Kuwait averaged 70.7 MIGPD, with a maximum that exceeded 90.2 MIGPD, during summer months[4]. Most of this water was used for irrigation of private and public gardens and landscaping, and a quantity was used for blending with distilled water, livestock feeding and construction work.

Kuwait has decided to rely on its' brackish water resource to transform part of it into potable water to be used in emergency cases by applying RO technology. In 1987, 13 RO units were installed and put into operation. The capacity of each unit is 0.250 MIGPD. These units are located in important places such as hospitals. Twenty more similar units are installed and operated on the sites of water reservoirs and pumping stations in different places around Kuwait. The total installed capacity production of freshwater by desalinating brackish water using the RO technique is 8.25 MIGPD[7].

The RO unit is erected inside three standard containers. The first container contains the membranes, high-pressure pumps, cartridge filter, flushing/cleaning tank, transfer pump, dosing stations, control panel and electrical switchgear. The second container contains two dual-media filters, feed pump, backwash air blower and associated pipes and valves. The third container, called the storage container, consists of chemicals required for six months continuous operation and two years of spare parts.

Brackish water is supplied to the feed water tank (capacity 50,000 IG) through the existing brackish water network/pipelines. The pre-treatment consists of dual-media filters and cartridge filters (5 μm size). Sulphuric acid (5 ppm) and antiscalant Flocon 100 (6 ppm) are dosed prior to the cartridge filters. Sodium bisulphite (2 ppm) is added at the suction of the feed pump. After the cartridge filter, the high-pressure pump raises the feedwater pressure to an operating pressure of 15 to 25 bar, depending on the feed operational conditions.

The RO section consists of eight pressure vessels operating in parallel, each containing six Filmtec, low-pressure spiral wound BW-8040 thin film composite membranes of 8 inch size. A product pump transfers the product water from the product tank to the existing freshwater networks. Brine from the unit is disposed of to the sea through the stormwater drainage system.

2.4 Multistage Flash (MSF) Plants

Thermal distillation depends on changing the physical conditions so that the water is changed from liquid phase into vapour phase leaving the salt in the brine stream and then the vapour condenses to the liquid phase resulting in distilled water. The distillation units in Kuwait use the MSF evaporation method. Each distillation unit consists of between 24 to 26 stages and the capacity of the units is between 5 to 6 MIGPD for each unit according to each station. There are six main stations in Kuwait with a total installed capacity of 234 MIGPD at normal temperature (90°C) operation, which could be raised to 256.8 MIGPD at high temperature (110°C) operation as presented in Table 3.

Table 3. *Total Installed Capacity of the MSF Plants in Kuwait*

Station	No. of Units	Unit capacity	Total Capacity
Shuwaikh	3	6	18
Shuaiba	6	5	30
Doha East	7	6	42
Doha West	16	6	96
Az-Zour South	8	6	48
Az-Zour South*	4	6	24
Total	44		258

*Under Construction

Fresh water gross production rose from 1773 MIGPY in the late 1950s to 73306 MIG in 1997[4]. The distillate water produced from the thermal station usually contains 3 to 30 mg/l salt depending on the design and effectiveness of the distiller. To produce freshwater, distillate water is being mixed with brackish underground water. This operation takes place in the blending station where the distillate water is mixed with brackish water at a ratio of 10:1. Also, disinfecting of the freshwater is being done by injecting a chlorine solution to kill bacteria and harmful organisms in the water, then caustic soda solution is added to maintain the pH value of the water within the required limits according to the World Health Organisation (WHO)[8].

3 SUMMARY

In summary, freshwater in Kuwait obtained either by RO or by MSF shows medium ranges for most mineral contents compared with bottled natural mineral water. Also, all freshwater sources in Kuwait (95% produced by MSF) are within the acceptable ranges of the WHO guideline values for drinking water (Table 4).

The collected data presented in Table 4 show that the bottled natural mineral water is not better than the potable water supplied by the Ministry of Electricity and Water (MEW) in Kuwait from the health point of view.

Table 4. *Chemical Analysis of Different Drinking Water Sources in Kuwait*

Elements	Undergro und	RO		MSF	Averag e	WHO	
	Freshwat er	Brackish water	Seawat er	Seawat er	Bottled	Max	Rec.
Bicarbonate (ppm)	155.00	16.8	4.9	57.2	109.7	500	100
Chloride (ppm)	9.00	21.0	187	75.5	26.75	600	200
Sulphate (ppm)	22.00	9.0	3.7	85.1	27.6	400	200
Nitrate (ppm)	8.0	3.0		1.03	5.11	450	10
Calcium (ppm)	45.0	0.7	0.7	42.8	26.9	200	75
Magnesium (ppm)	7.0	0.5	1.8	9.3	10.5	150	30
Potassium (ppm)	4.0	0.5	3.3	2.1	2.5	12	10
Sodium (ppm)	9.0	19.0	128	44.1	21.6	200	20
pH	7.8	7.3	6.5	7.84	7.47	9.2	7.0-8.5
TDS (ppm)		74.0	375	322.8	180.62	1500	500

Finally, careful selection should be made of drinking water, taking into consideration the following:

- Low values of sodium content for hypertension and heart weakness.
- Calcium concentration contributes in building bones and teeth.
- The body finds it difficult to deal with excess potassium, resulting in kidney stress.
- It is better to keep the water pH value in the rage of 6 to 8. A pH value below 5 makes the water aggressive and can lead to problems such as leaching minerals from teeth if used regularly.
- The higher the TDS content in the water, the saltier it will become.

4 ACKNOWLEDGMENT

The authors wish to acknowledge the financial support and co-operation of the Kuwait Foundation for the Advancement of Sciences (KFAS).

References

1. T. Dabbagh, 1995, World Congress on Desalination and Water Science, Abu Dhabi, U.A.E, November 18-24.
2. G.A. Le Moigne, A. Subramanian, M. Xie, and S. Gilter, 1994, World Bank Technical Report 263.
3. M. Sahlawi, 1999, WSTA 4th Gulf Water Conference, Feb. 13-17, State of Bahrain, pp. 37-44.

4. Statistical Year Book, 1998, Ministry of Electricity and Water, State of Kuwait
5. B. Al-Nashi and J.G. Anderson, 1997, Conference Proceedings Vol.2, The Third Gulf Water Conference, Sultanate of Oman, 8-13 March, pp. 677-691.
6. Al-Rawdatain Water Bottling Company, 1995, Annual Report, Kuwait.
7. A. Malik, N. Younan, B. Raq and K. Mousa, 1989, Fourth World Congress On Desalination and Water Reuse, Kuwait, November 3-8, pp.341-361.
8. K. Al-Fraij, M. El-Aleem and H. Al-Ajmi, 1999, WSTA 4th Gulf Water Conference, State of Bahrain, Feb.13-17, pp.823-840.

NANOFILTRATION FOR COLOUR REMOVAL – 7 YEARS OPERATIONAL EXPERIENCE IN SCOTLAND

E. Irvine

West of Scotland Water,
Engineering Services,
John MacDonald House,
296 Vincent Street,
Glasgow
G2 5RG

A.B.F. Grose

Thames Water Research & Technology,
Spencer House Manor Farm Road,
Reading
RG2 OJN

D. Welch

PCI Leopold Ltd.,
1 Kilduskland Road,
Ardrishaig.
Argyll
PA30 8EH

A.Donn

PCI Leopold Systems Ltd
Laverstoke Mill,
Whitcurch,
Hampshire
RG28 7NR UK

1 INTRODUCTION

In Scotland drinking water is treated and supplied by three Water Authorities, namely East, North & West of Scotland Water. Most of the population is concentrated in the central belt where a network of Treatment Works and pipelines supply drinking water to the customers. The other area of the country is sparsely populated with many, small, remote communities. Typically each community is supplied from a local treatment works, owned and operated by the Water Authority. The majority of these treatment works supply less than 1000 people. Operators have to overcome the difficulties and costs associated with the operation and maintenance of a large number of small, isolated treatment plants, all of which must meet the same water quality standards as the larger, more accessible works.

More than 50 % of Scotland's water is supplied form surface water from either "burns" (small streams) or lochs (lakes). The terrain is often mountainous with much of the higher land afforested or used as rough moorland grazing for sheep, deer and cattle. These upland waters are very soft, exhibiting low mineralisation, but contain high concentrations of natural organic matter (NOM), iron and manganese. The colour is imparted by organic substances, predominantly humic and fulvic acids which upon chlorination result in the formation of disinfection by products (DBPs) in the form of trihalomethanes (THMs). The treatment of these waters is further compounded by the flashy nature of the raw water with the existing rudimentary treatment being ineffective and incapable of allowing compliance with the water quality standards.

Drinking water standards are laid down in The Water Supply (Water Quality) (Scotland) Regulations, to comply with EC Directives are similar to those in the rest of the UK. Historically, on these rural works employing crude treatment, the two most common water quality statutory failures are Bacteriological and THM's.

With the ever more restrictive drinking water quality standards, expected operational practices – Badenoch and Bouchier reports allied with higher customer expectations and demands all impose on the Authorities the need for a reliable and robust treatment process.

Lochgair WTW is a works that is typical of all the above. It was proposed that the old treatment works at Lochgair could be replaced with a nanofiltration (NF) membrane plant. This would be capable of removing both the micro-organisms detected in the source water, the high iron concentrations and the high colour that acts as a precursor for THM formation.

2 PROCESS DEVELOPMENT IN SCOTLAND

When West of Scotland Water Authority (WOSWA) needed to up-grade the existing treatment works at Lochgair, they were faced with a number of issues, all typical of many treatment works through out Scotland:-

- ◆ The old treatment works (upward slow sand filtration, pH correction and final chlorination), was no longer capable of producing water of sufficient quality to achieve regulation compliance
- ◆ Derogations allowing temporary exclusion for certain parameters were in place at Lochgair for colour, iron and THMs, however these were due to expire thereby requiring future full compliance of all parameters.
- ◆ Treatment works are frequently very small, typical daily demand at Lochgair was between 20 -30 m^3/d.
- ◆ The site is very remote, reached by a rough dirt track up the side of a mountain
- ◆ Raw water quality showed considerable variation and was subject to "flash" conditions when the quality could change extremely rapidly.
- ◆ Resources were limited, the site was visited on a weekly basis.

In 1991 a six element spiral NF pilot plant using Magnum 8231LP cellulose triacetate membranes supplied by Fluid Systems was installed. Personnel from WOSWA, Thames Water Research & Technology and PCI Membrane Systems Ltd carried out an extended programme of trials at the site.

The objectives of these trials were:-

- ◆ to develop a nanofiltration membrane process to remove
  - - colour and other natural organics
  - - pathogenic organisms
- ◆ to demonstrate that the process operation was
  - - reliable and automatically controlled at a remote site
  - - able to treat "flashy" raw water quality
  - - full compliant with water quality regulations
- ◆ to optimise cleaning and sanitisation procedures

♦ to optimise pre- and post treatment requirements
♦ to identify a viable waste disposal route
♦ to demonstrate water quality improvements at the customers tap.

2.1 Spiral Wound Membrane Nanofiltration

The successful pilot trials led to the installation in 1992 of a 70 m^3/d membrane plant at Lochgair. Leaching and bacterial growth trials were carried out in order to achieve the required UK Government materials approval.

Raw water was gravitated down to the plant from a simple weir intake at the loch, and was subjected to pre-treatment. Initially the pre-treatment was achieved using the existing sand filters, however they required regular operator attendance and latterly a dual media filter with automatic backwash was installed. The pre-treated water passes through a 5 um cartridge filter and then on to the membranes operating at an inlet pressure of approximately 110 psi (7.6 Bar). The NF membranes remove the colour from the water with the resultant reduction in THM formation upon final chlorination. The membrane product water was aggressive and required pH corrected by dosing sodium carbonate to control corrosivity.

Table 1. *Raw water quality a Lochgair WTW*

Parameter	Units	Raw Water
Colour	°Hazen	73
TOC	mg/l	8.4
pH		6.21
Turbidity	NTU	1.1
Conductivity	uS/cm	68
Total hardness	mg/l $CaCO_3$	5.4
Alkalinity	mg/l $CaCO_3$	9.6
Total aluminium	ug/l	71
Total iron	ug/l	673
Total manganese	ug/l	29
Coliforms	per 100 ml	21
E.coli	per 100 ml	11
Total THMs	ug/l	0

Table 1 gives a typical raw water quality for the Lochgair. The total organic carbon (TOC) concentration (9.61 mg/l) is predominantly a result of the naturally occurring organic compounds that give rise to the high colour (89 ° Hazen). The high raw water total iron concentration (673 ug/l) is typical of these waters where the organic fraction readily becomes complexed with metals such as iron and manganese.

The membranes required chemical cleaning at regular intervals to control fouling. The main causes of fouling were organic colour complexed with iron and biological film formation. The cleaning programme alternated neutral enzymic detergents and weak acidic solutions. Sanitisation with sodium hypochlorite was also carried out periodically.

The use of the spiral wound membrane configuration at Lochgair proved that soft coloured waters could be satisfactorily treated by membrane nanofiltration. As a result a second spiral wound membrane plant has been installed at Bunessan on the island of Mull. Due the large tourist interest the plant has a variable flow with a maximum 550 m^3/d.

Despite the success of these membrane plants, spiral wound membranes were perceived to have the following disadvantages:

♦ The requirement for regular chemically cleaning means that chemicals had to transported and stored on site.
♦ The spent chemical clean fluid required an acceptable disposal route.
♦ Efficient pre-treatment of the raw water was required to prevent the spiral wound modules becoming "plugged" with particulates.

2.2 Tubular Membrane Development

In order to address the issues mentioned above, the joint development team carried out trials on 12 mm diameter tubular CA membranes supplied by PCI. These tubular membranes proved to possess a number of advantages over the spiral wound membranes:-

♦ No pre-treatment is required. High turbidity waters can be treated with out the need for pre-filtration.
♦ The membranes could be cleaned mechanically. Foam balls are passed automatically up the tube to clean the membrane surface.
♦ Chemical cleaning is no longer a regular requirement and can be limited to annual or biannual cleaning.
♦ The chemical clean can be tankered off site to avoid associated disposal issues.

Successful development led to the first tubular installation at Cladich WTW in 1994. Similar water quality problems to Lochgair exist with highly coloured flashy raw water and subsequent high THM formation potential, together with a risk of contamination by *Cryptosporidium*.

Prior to the installation of the membrane plant, the site required a daily visit by the operator though despite this attention, the final water was subject to microbial failures, periodically requiring customers to boil the water before use. Since the installation of the membrane plant the water quality now achieves full regulation compliance. Table 2 provides typical analytical results achieved at Cladich WTW.

The Cladich membrane installation has a capacity of 20 m^3/d, the plant was sized to meet peak flows, however normal demand is well below this and it typically operates for 3-4 hours a day. The product water is dosed with sodium carbonate and sodium hypochlorite is dosed to give a chlorine residual. Plant operation and cleaning is carried out automatically. The site visit frequency has been reduced from daily to once per week to collect statutory samples and replenish chemicals used for final water treatment. Subsequent development work has replaced the sodium carbonate dosing system with a robust limestone contactor.

Table 2. *Water Quality Results - Cladich*

Determinand	Raw Water	Product Water	Post Reservoir	Distribution
Colour (°H)	80	0	0	2
TOC (mg/l)	2.3	0	0.7	0.8
Turbidity (NTU)	0.7	0	0	0
Conductivity (uS/cm)	42	27	41	59
pH	7	7.2	8.3	8.5
Total THM (ug/l)	0	0	12	19
Total Coliforms /100 ml	15	0	0	0
Ecoli /100 ml	7	0	0	0
Colony Count 1 day @ 37°C /ml	45	0	1	2
Colony Count 3 day @ 22°C /ml	190	38	0	1

Based on the above success a number of other B1 plants were constructed at Kirkmicheal, Tomnavoulin for North of Scotland Water, Balquhidder for East of Scotland Water, Ballygrant, Carrick Castle, Dervaig for West of Scotland Water and Eredine for Argyll & Bute District Council.

2.3 C10 Tubular Membrane Module

The initial design of tubular membrane plants utilised PCI standard stainless steel B1 module design. In order to reduce construction costs significant development has gone into producing a tubular membrane module that can compete with spiral wound designs for the targeted plant size range of up to approximately 3.8 Mld.

This module, known as the C10, contains over 4 times the membrane area of the B1 modules and is manufactured in ABS plastic. The plant is skid mounted and is an extremely compact plant.

The first C10 plant treating 440 m^3/d, has been installed at Gorthleck in May 1998 for the North of Scotland Water Authority. North of Scotland Water Authority capital programme allows further nanofiltration plants to be constructed. Three nanofiltration plants are currently under construction or have been recently commissioned for the East of Scotland Water Authority. East of Scotland Water Authority capital programme allows further nanofiltration plants to be constructed. West of Scotland Water Authority have entered into a framework agreement with PCI Leopold to supply nanofiltration plants.

In total, more than twenty PCI Leopold nanofiltration plants are operating in Scotland.

Overseas, interest in the Fyne treatment process continues to grow with the recent order for a plant at Chapel Island, Canada being one of the highlights.

3 DRINKING WATER QUALITY

Following the installation of the nanofiltration plant the improvement to the quality of the drinking water is significant. In addition to the information displayed in Table 2, Table 3 gives typical results for final water from treatment works before and after commissioning of the nanofiltration plant.

Table 3. *Water Quality Results*

Determinand	Final Water(1)	Final Water(2)	Final Water(3)	Final Water(4)
Colour (°H)	31	3.0	25	3.0
Iron	289	< 10	451	17
Manganese	26	< 5	11	23
Turbidity (NTU)	1.0	0.27	4.3	0.2
Conductivity (uS/cm)	184	154	199	60
Total THM (ug/l)	388	25	149	< 10
pH	7.63	8.65	8.5	9.28
Total Coliforms /100 ml	0	0	30	0
Ecoli /100 ml	0	0	5	0
Colony Count 1 day @ 37°C /ml	14(15)	3 (8)	20 (44)	4 (6)

Columns headed (1) & (2) relate Bunessan WTW, a spiral plant, and Columns headed (3) & (4) relate to Carrick Castle, a tubular plant. Columns headed (1) & (3) are before commissioning and Columns headed (2) & (4) are after commissioning of the NF plant. The colony count is the number of occurrences in a six month period with the maximum count bracketed. All figures are maximums and are taken from water as it enters the distribution network.

4 COST COMPARISIONS

The membrane treatment plants are designed to provide at least cost, a reliable treatment process at remote treatment works, which require little or no operator input. Operator input is required to monitor the plant, batch-up chemicals, take regular samples and on occasion chemically clean the membranes. As operational experience has been expanded the membrane life guarantees are being extended from the initial 1 year period to 3 years. Some membrane plants have benefited from membrane life lasting longer than the increased guaranteed life.

Significant differences to duties performed by the operator are required on a membrane plant when compared to the traditional coagulation chemistry based plants and often the biggest is the reduction in the number of visits that are required. This is important when operators spend time travelling between treatment works. Table 4 illustrates the differing duties for different treatment plants, including the different cleaning requirements for the tubular and spiral membrane plants. An added benefit is that overhead costs are reduced because the membrane process is more reliable, the number of water quality failures have been reduced, allowing all staff to go about there normal duties rather than respond to emergencies.

Table 4. *Comparison of Operators Duties*

Treatment Works	Treatment Process	Operator Duties	Hours Average Per Day
Ardrishaig	Dissolved air flotation and two-stage filtration. Sludge thickener & plate press	1. Water quality checks 2. Floc test 3. Chemical batching 4. Sludge plant 5. Instrument calibration 6. Housekeeping	1.0 0.5 1.0 0.5 0.5 0.5
Tighnabruaich	Pressure sedimentation and two stage filtration	1. Water quality checks 2. Floc test 3. Chemical batching 4. Filter backwash 5. Instrument calibration 6. Housekeeping	1.0 0.5 1.0 2.5 0.25 0.25
Bunessan	Spiral Membrane plant	1. Chemical batching & clean 2. Housekeeping	1.0 0.05
Carrick Castle	Tubular Membrane plant	1. Chemical batching & clean 2. Houskeeping	0.05 0.05

Table 5 illustrates the operational cost benefits of membrane plants compared to other treatment processes. Ardrishaig WTW is a modern fully automated plant having been constructed some six years ago. Tighnabruaich WTW is an older installation relying on coagulation chemistry with the pressure filter plant manned for a full day, every day. Water quality failures have been identified and the WTW is due for capital investment.

The major benefit of operating a membrane plant to the Water Authority is the reliability and robustness of the process. As highlighted in Table 5 the operational cost of producing water in rural areas is not cheap.

When compared to the conventional coagulation chemistry process the NF process offers substantial saving to chemical costs, substantial saving in operator attendance, saving in waste disposal costs and because less equipment is required maintenance costs are reduced. The cost disadvantage is the increase in power consumption and the membrane replacement cost. The latter can be spread evenly over the years by service agreements.

The costs shown in Table 5 also illustrate the cost of operating small, local treatment works located remotely, in rural locations. The West of Scotland Water meter rate is 45.6p/m cu.

Table 5. *Operational Cost Comparison*

	Ardrishaig	Tighnabruaich	Bunessan	Carrick Castle
Design Flow	3,200	600	550	70
Operator	15,045	17,938	2,653	1,625
Replacement	10,210	6,000	N/A	N/A
Membrane Replacement	N/A	N/A	2,880	2,500
Maintenance	16,850	4,170	2,830	150
Waste Removal	4,650	3,100	-	-
Power	15,334	420	4,710	5,496
Chemicals	38,750	12,850	3,740	200
Total	**100,839**	**44,478**	**16,813**	**9,971**
Cost p/m cu	**3,151**	**7,413**	**3,057**	**14,244**

Notes
1) Bunessan chemical costs include for delivery to the island.
2) Bunessan & Carrick Castle replacement & maintenance costs are related to equipment other than the membrane associated with the systems.
3) All costs are in £ UK.

5 CONCLUSIONS

The Fyne treatment process has been developed with a substantial number of plants are now operational in Scotland. Compliance with water quality regulations has been demonstrated. The customers expectations of the water appearing at their tap has been increased, particularly the non-use of chemicals is recognised as an advantage.

Nanofiltration has proved to be cost effective for small treatment works and tubular membranes have proved to be ideally suited as routine mechanical cleaning can automated, thereby reducing operators attendance and costs. Although a number of options are available to the Water Authority waste disposal issues are site specific and require consent from SEPA.

The use of membranes is now accepted as a water treatment option in Scotland and the UK. The use membrane treatment process is becoming more widespread and the process can offer advantages at larger treatment works. The technology continues to be developed which should bring further cost benefits to the customer.

Acknowledgements

The views expressed in this paper are those of the authors and not necessarily those of their respective organisations. The authors wish to acknowledge the assistance received from plant operators at West of Scotland Water in producing this paper.

ULTRAFILTRATION FOR 90 MLD *CRYPTOSPORIDIUM* AND *GIARDIA*-FREE DRINKING WATER.
A CASE STUDY FOR THE YORKSHIRE WATER KELDGATE PLANT

F.N.M. Knops

NORIT Membrane Technology
P.O. Box 89
7550 AB Hengelo
The Netherlands

B. Franklin

Earth Tech Engineering
Wentworth Business Park
Tankersley,
Barnsley
South Yorkshire S75 3DL
United Kingdom

1 INTRODUCTION

Cryptosporidium and other chlorine resistant micro-organisms are a problem for potable water supply companies throughout the world. Numerous problems have been reported with the *Cryptosporidium* outbreak in Milwaukee (USA) which occurred in April 1993 as most notorious example.

Cryptosporidium and *Giardia* form cysts which are very resistant to commonly used disinfection methods, e.g. a UV dose of 100 mJ/cm^2 will only achieve a 90% kill rate. Note that this dose is about twice the dose normally being used for drinking water disinfecting.

The cysts are quite large, relative to other microorganisms, the *cryptosporidium* oocysts measuring roughly 2 to 10 micrometers in diameter. This is an advantage from a water treatment point of view as they should be easy to remove with an absolute filter. NORIT Membrane Technology's ultrafiltration membrane provides this absolute filter.

2 NORIT MEMBRANE TECHNOLOGY

NORIT historically has been a company deeply involved in water treatment through its activated carbon division. NORIT has seen the importance of additional water treatment needed by its customers. Therefore NORIT started four years ago to form a separate division, NORIT Process Technology. This division consists of several companies active in water treatment and in engineering and manufacturing of process equipment.

NORIT Membrane Technology was founded four years ago as a company focussed on developing membrane applications for the following markets:
- Breweries. Microfiltration shows a promise in replacing Diatomaceous Earth precoat filters for yeast removal.
- Industrial process and cooling water production. Surface water and effluents from wastewater treatment plants will be increasingly used for process and cooling water instead of potable water. Both water sources need to be properly pre-treated:

ultrafiltration is the key for low cost and reliable production of process and cooling water.

♦ Pre-treatment for seawater desalination. Conventional pre-treatment can often not remove contaminants, such as silt and algae, which interfere with reverse osmosis membrane operation. Ultrafiltration can achieve this in a single step, without the need for dosing chemicals, such as chlorine or flocculants.

♦ Potable water production. Increased awareness of water borne diseases has shown that conventional treatment has its drawbacks. Ultrafiltration will provide an absolute barrier.

Four years ago NORIT Membrane Technology became the first company within the Process Technology division. Since then a membrane manufacturing facility has been added to this division as well. This gives NORIT Membrane Technology the opportunity for in house development of membranes as well as applications.

3   PROJECT DESCRIPTION AND BACKGROUND

Yorkshire Water Services, one of the large private water companies in England, has a potential problem with the ground water sources used to supply the city of Hull and the surrounding areas. For the vast majority of time the water is of excellent quality but there are occasional spikes of turbidity after heavy rainfall. Although *cryptosporidium* has not been detected the sources are potentially at risk.

Pilot plant work was carried out which indicated that ultrafiltration would be a cost effective method of treating the water to provide the required barrier. Yorkshire Water decided to proceed with a scheme to treat the four affected sources at a central 90 Ml/day ultrafiltration plant.

Earth Tech Engineering was appointed as the main contractor to refurbish the borehole sources and construct the new works. Earth Tech, part of the TYCO group of companies, subsequently went out to competitive tender for a membrane plant, capable of achieving a minimum of 4 log removal (99.99%) of *cryptosporidium* sized particles. NORIT Membrane Technology was awarded the subcontract to design and supply the membrane plant.

4   FULL SCALE PLANT

4.1 Design

NORIT Membrane Technology used the following criteria for engineering an ultrafiltration system suitable for the Keldgate project:

♦ Reliability. This is of foremost importance, as the installation will be providing drinking water to a large area. Plant down time has to be avoided; therefore NORIT Membrane Technology used a design which has no single point of failure.

♦ Flexibility. The flow through the plant will vary from 30 to 90 Ml/day. The plant uses a modular design, which enables part of the plant to be switched off in the case of low demand. Future expansion up to 99 MLD can be realised by adding modules to the plant.

- Quality control. The water produced by the plant has to be of a proven quality. On line monitoring and long term logging of the water quality will ensure a reliable plant operation and supply.
- Ease of operation. Unlike conventional water treatment plants, an ultrafiltration system can not be operated manually. All flows and timers have to be tightly controlled; therefore full automatic operation is implemented.
- Minimisation of the environmental impact. Production of wastewater, energy consumption and chemicals usage has to be minimised.

Keeping the above criteria in mind, NORIT Membrane Technology designed the following ultrafiltration system:

1. The ultrafiltration plant consists of eleven identical modular membrane racks. Each membrane rack can hold up to 3360 m^2 of membrane area.
2. During normal operation up to nine membrane racks will be used to treat raw water. The capacity can be varied either by controlling the flow to the individual racks or by switching racks off line. These nine racks in production are called the primary ultrafiltration system.

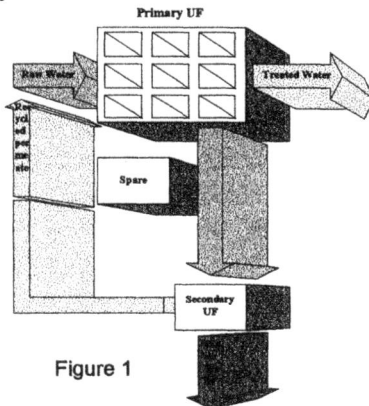

Figure 1

3. The above racks will generate approximately 5% wastewater. One rack is used as a secondary ultrafiltration system to reduce this volume. The filtrate from this rack will be returned upstream of the primary ultrafiltration, see figure 1.

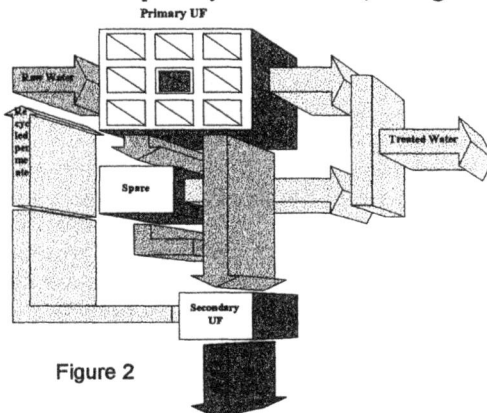

Figure 2

4. The eleventh membrane rack is a common spare for both the primary and the secondary ultrafiltration systems. If one of the primary membrane racks is out of service the spare will be used as primary ultrafiltration system (figure 2), if the secondary ultrafiltration is out of service the spare will be used as secondary ultrafiltration (figure 3).
5. Four independent PLC's will control up to three membrane racks each. This means no single point of failure and allows for future expansion by adding a further membrane rack.
6. The PLC's will control normal operation of the ultrafiltration racks. A Supervisory Data Acquisition and Control System (SCADA) is used as operator interface and for telemetry. In addition to the SCADA system each PLC will be equipped with a local control panel. This control panel will be used for local read out of the instrumentation and for operation in case the SCADA system is out of service.

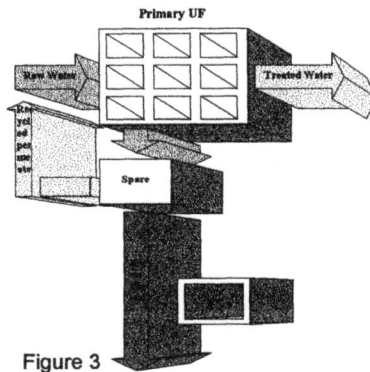

Figure 3

4.2 Primary Ultrafiltration

The primary ultrafiltration system uses membranes with a pore size of 150,000 Dalton. This is roughly equivalent to 0.05 micrometer, one hundred times smaller than the size of the *Cryptosporidium* oocyst. The membrane will not only remove *Cryptosporidium* but also viruses. The membranes are formed in the shape of hollow fibres, with an internal diameter of 0.8 millimetres. Ten thousand membranes are bundles in a membrane module; this module holds 35 m^2 of membrane area. The module has a diameter of 8 inch and a length of 60 inch, identical to a RO Magnum module. Four modules will be fitted in one pressure vessel 6 meters long.

Each of the nine primary membrane racks holds twenty-four pressure vessels in four stacks of six pressure vessels (figure 4). Membrane cleaning will be done either by a flush with clean water (backwash) or by a flush with low concentrations of chemicals (Chemically Enhanced Backwash). A Chemically Enhanced Backwash does not require circulation of cleaning chemicals or elevated temperatures, but is rather performed by a simple wash followed by an extended soak time. This minimises chemical usage and energy consumption. Three Cleaning in Place units will be used: each unit will clean three membrane racks.

Figure 4

4.3 Secondary Ultrafiltration

The secondary ultrafiltration plant uses the same membranes as the primary ultrafiltration plant. The rack is identical to the primary ultrafiltration racks.

The concentrate from the primary ultrafiltration is buffered in a small tank being fed to the secondary system. This can be done in two ways: either start/stop operation on high and low level in the tank, or by controlling the flow and matching secondary ultrafiltration capacity to waste production of the primary plant.

The secondary ultrafiltration will be cleaned in the same way as the primary ultrafiltration. The frequency of cleaning and the concentration of the chemicals to be used can be set independent of the setting of the primary ultrafiltration. The CIP unit is common for the secondary ultrafiltration and for the spare membrane rack.

The washwater generated by cleaning the secondary ultrafiltration is held in a buffertank where it is neutralised before discharge to sewer. The maximum discharge rate is 15 litres per second.

4.4 Membrane Integrity Testing

Both the primary and the secondary ultrafiltration system use NORIT's SIM®-test. This test (Spiked Integrity Monitoring) is a new design specifically developed for membrane integrity control at potable water plants.

For potable water uses, the feed water is often of high quality and the purpose of the membrane is to remove micro-organisms. The customer demands a high degree of membrane integrity and defects must be detected quickly, preferably by an on line test, which can be carried out at frequent intervals.

Conventional water quality instruments, such as turbidity monitors, are not suitable for this. Defects would not be detected unless there were multiple failures. Moreover this would not become apparent until turbidity was present in the raw water.

Historically membrane defects are detected by using a bubble point test: this test uses the fact that wetted membrane pores do not pass air. Air passage through a membrane indicates a defect. This can be done in two different ways: either by applying a vacuum to the permeate side of the membrane or by applying an overpressure to the feed side of the membrane. The vacuum or pressure decay can be measured and indicates a defect. As an alternative the flow through the membranes can be monitored.

These test procedures have the following shortcomings:
- ◆ The ultrafiltration process has to be interrupted. This test can not be done on line and thus lowers net plant capacity.
- ◆ This test is very sensitive to outside disturbances: even minor outside leaks like e.g. incorrectly positioned valve seats will give a false impression.
- ◆ There is no direct relationship between the readings from the test and the actual removal rate for micro-organisms by the membranes.
- ◆ The maximum membrane area to be simultaneously monitored is limited. This effect is caused by the diffusive airflow through the pores of the membranes.

NORIT Membrane Technology has invented a new integrity monitoring system that overcomes all above problems: it is called Spiked Integrity Monitoring.

The SIM®-test is based on the principle that the ultrafiltration membrane is challenged with a known dose of *Cryptosporidium* sized particles and the permeate quality is on line monitored. The particles to be used for spiking the feed water are made of powdered activated carbon. This activated carbon dust is harmless and has DWI approval.

If a known, high dose of particles is fed to the membranes, the amount of particles in the permeate can be used for calculating the log removal for *Cryptosporidium* sized particles. The particles present in the feed water give a negligible contribution to the overall particle counts and to the log removal readings of the SIM®-test.

NORIT's SIM®-test uses specially designed high accuracy particle monitors and powdered activated carbon with a very accurate particle size distribution and a mean particle size close to the size of *Cryptosporidium*.

NORIT's SIM®-test is characterised by the following advantages:
- ◆ The test can be done on line. No production losses will occur. The test can be performed as often as one wishes.
- ◆ The actual log removal rate will be recorded and can be used for future references. It will be possible to plan maintenance in advance, based on anticipated removal rates.
- ◆ The SIM®-test is independent of membrane area to be monitored. The activated carbon will only pass through defects, no diffusion will occur.

For the Yorkshire Water Keldgate plant NORIT Membrane Technology will use four independent SIM®-systems.

4.5 Layout of the Plant

As described above the ultrafiltration plant consists of eleven membrane racks of up to twenty-four pressure vessels each. Each one of these membrane racks has its own feed pump and its own prefilter. The prefilter is used as a safety filter to protect the down stream membranes from coarse particles. The membrane racks (figure 4) are mounted on stainless steel frames. These will be pre-assembled and tested prior to shipment.

The Keldgate site consists of three separate buildings: the main building, the chemicals building and the control building. The control building will house the SCADA system. All local automation equipment (PLC's) will be located on the membrane racks in the main building.

The main building is designed with two floors: the basement and the ground floor. The basement holds the ancillary equipment, such as the pumps, the pre-filters and the buffertanks. The membrane racks are positioned on the ground floor and are connected through the floor to the ancillary equipment (figure 5).

5 CONCLUSION

At the time of order award the Keldgate plant was the largest municipal ultrafiltration plant in the world. This plant will be a reference project for even larger plants to be built in the near future.

The following data characterise the Keldgate plant:

◆ Design capacity   :   90 MLD equals 1,000 litres per second.
◆ Maximum hydraulic capacity   :   1,250 litres per second.
◆ Total installed membrane area   :   37,000 m^2.
◆ Total membrane fibre length   :   15,800 kilometres.
◆ Power consumption   :   500 kW.

The complete project will be commissioned mid 2000. This means a lead-time of just over twelve months. If one takes the size and the innovations realised within this project into account, this is a huge achievement, only possible due to a joint effort of enduser, engineer and membrane supplier.

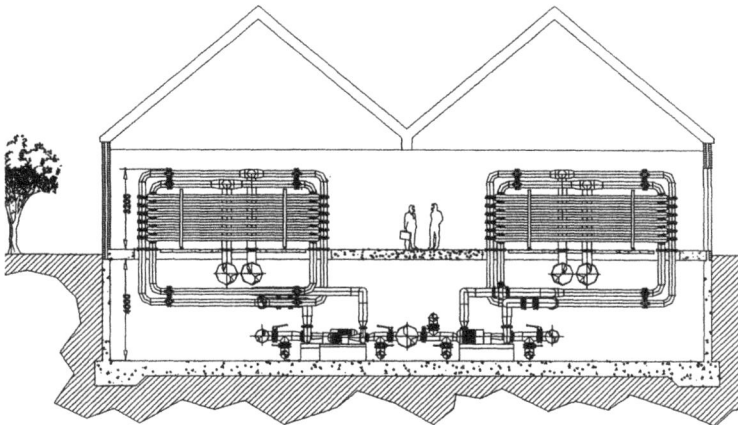

Figure 5

6 ACKNOWLEDGEMENTS

The authors wish to thank Mr. John Lever and Mr. Norman Johnson of Earth Tech Engineering, Mrs. Rosemary Smith of Yorkshire Water Services, Mr. Peer Kamp of Provinciale Waterleidingmaatschappij Noordholland, Mr. Stefan van Hoof of NORIT Membrane Technology and Mr. Paul van Oort of NORIT Process Technology for their contributions to this paper.

APPLICATION OF A NEW GENERATION MICROFILTRATION PROCESS FOR LARGE SCALE WATER AND WASTEWATER TREATMENT

Warren T. Johnson, Project Manager – Product Development
Alex Patterson, Senior Development Engineer

USF Memcor Research Pty Limited,
Sydney,
Australia

1   ABSTRACT

A new generation of USF Memcor's microfiltration (CMF) process has been developed. The design utilises submerged membranes in an open tank rather than multiple pressure vessels, as in the existing CMF technology.   The technology is known as CMF-S (submerged).   It is the result of several years of R&D and represents a natural progression of the CMF technology into very large-scale filtration applications. The simplification of the process and system makes it cost effective even at capacities greater than 100 ML/day, using unit multiples of approximately 20 ML/day.

This paper briefly reviews the history of CMF development and describes the first commercial project using the new CMF-S in Marulan, Australia.   This small scale system helped to develop the design concepts used for the design of the first large scale CMF-S plant in Bendigo, Australia, due to be commissioned in late 2000. This project includes four water treatment plants, the largest of which is 126ML/day.

2   INTRODUCTION

Low pressure membrane systems have come a long way in recent years. Tightening water quality regulations, particularly with respect to Cryptosporidium and Giardia, has fueled interest in membrane technologies. As a result there has been a rapid growth in the membrane market, which has seen the technology develop and expand along with it. As little as 5 years ago systems that were amongst the largest installed are little more than pilot units on today's scale. Membrane units are installed or under construction for systems over 100 ML/day, and projects up to 500 ML/day are seriously considering microfiltration over conventional treatment methods.

However, existing membrane systems generally struggle to be cost effective against conventional processes on large scale applications (>100 ML/day). Unit sizes tend to be relatively small (typically <5 ML/day) limiting the usual economies of scale. A new version of CMF using submerged membranes (CMF-S) was specifically developed with the objectives of:

- simplifying the design to reduce capital cost
- to scale-up the microfiltration process to plants >100 ML/day.

This paper briefly reviews the history of CMF development. It describes the first small scale commercial project using CMF-S at Marulan in Australia and the successful bid for the AQUA 2000 project, the first large scale CMF-S plant and the largest microfiltration plant in the world – 126ML/day, due to be commissioned late 2000.

3   EVOLUTION OF FOURTH GENERATION CMF-S TECHNOLOGY

USF/Memcor introduced its CMF microfiltration technology in the early eighties. The first generation of systems were relatively expensive and were therefore limited to high value low volume process streams such as wine, fruit juice, and pharmaceutical applications. They were typically manufactured into small skid mounted units using standard plumbing fittings. Further cost reductions and simplifications were made in the second generation M1 and M2 series by using injection moulded manifolding and doubling the area of each module (by increasing the length). Microfiltration was now becoming cost effective for small-scale water and wastewater treatment and in 1991 the first municipal drinking water plant was installed in Tooborac, Victoria, Australia (170 kLday).

The third generation of microfiltration systems, the M10 and M10C series, were introduced in 1992 specifically to allow high volume processing of water and wastewater using repairable modules. A microfiltration block of this type contains 2 to 90 modules depending on the required filtration capacity. One 90-module block filters 2.5 to 4.5 ML/day depending on the feed water quality. Large plants duplicate the 90-module block.

The development of the M10/M10C technology coincided with the introduction of the surface water treatment rule in the US and recognition that membranes offered an elegant solution to the problem of removing chlorine tolerant organisms such as Crytosporidium and Giardia. This resulted in a rapid growth in CMF installations. Growing acceptance of the technology and further developments have seen system costs tumble (Figure 1). There are now over 100 Memcor CMF potable water plants with a capacity of between 1 and 126 ML/day installed or under construction, with a total MF capacity in excess of 500 ML/day.

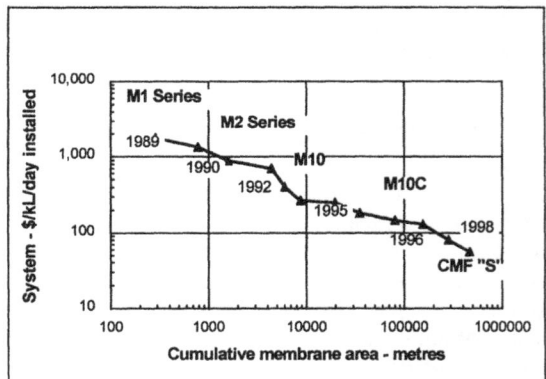

Figure 1 – *Cost reduction over time*

USF Memcor's most recent development is the CMF-S (submerged) series. This fourth generation system is the result of several years of R&D and represents a natural progression of the CMF technology into very large-scale filtration applications. The simplification of the process and system allows it to be applied cost effectively at capacities greater than 100 ML/day using unit multiples of approximately 20 ML/day.

For the first time this allows membrane filtration to compete on a cost par with conventional processes on even the largest scale filtration applications.

4 THE CMF-S PROCESS

In the CMF-S process the membranes are submerged in an open tank and filtrate is withdrawn from the modules under suction. The membranes operate in direct flow using the same nominal 0.2um polypropylene membrane as the conventional CMF process has employed for many years. A periodic backwash is required to dislodge the solids and chemical cleaning is employed at intervals to restore membrane performance.

4.1 CMF-S Backwash

Earlier CMF generations used a unique gas backwash to keep the membrane surface clean. The CMF-S backwash is based on the same principles with the exception that the air is applied externally to the fibre surface rather than through the membrane wall. The key to its success is a design that allows the air to be delivered evenly into the depths of a highly packed membrane module.

A short filtrate backwash supplements the scrubbing action of the air to help further dislodge solids. Solids dislodged from the membrane are drained from the tank at the end of the backwash cycle. This process ensures that there is no accumulation of solids within the membrane tank that can affect the membrane performance or the water quality. The result is a backwash process with efficiency similar to conventional CMF.

4.2 CMF-S Clean-In-Place (CIP)

Chemical cleaning is achieved pumping cleaning solution from the CIP tank to the cell containing the membrane modules. The membranes are contacted with the chemicals and periodically aerated. At the completion of the cycle the cleaning solution is recovered to the CIP tank. The membranes are then rinsed and returned to service.

4.3 CMF-S Integrity Monitoring and Control

Both CMF and CMF-S use three key steps to achieve control of system integrity – pressure decay testing, identifying leaks and isolating leaking sub-modules. Pressure decay testing allows the integrity of the barrier to be quantified without needing to rely on measurements of treated water quality. The test is also more sensitive to changes in integrity than any current water quality testing methods can achieve.

Air is applied to the filtrate side of the system at a pressure below the bubble point (typically 100 kPa). The air displaces filtrate from the lumen through the membrane wall. Once all the liquid has been expelled from the lumen the system is isolated and the pressure decay monitored. The rate of pressure decay is a direct measure of the membrane integrity and can be related to log removal of particles larger than the pore size (e.g. Cryptosporidium) (Hong et al.,1999).

If pressure decay testing indicates a drop in integrity the source of the leak is easily located by visual inspection over the top of the tank. Groups of modules can be isolated for later repair if required.

5   APPLICATION OF CMF-S AT MARULAN, AUSTRALIA

5.1   Background to the Project

The Marulan project is the first commercial installation of the CMF -S technology. The town of Marulan is located 20 km east of Goulburn and has a population of 650 people. The town's water supply system was installed initially in 1967 at a capacity of 1.1 ML/day. It comprised reticulation and chlorine disinfection but no water treatment.

In order to meet the multiple objectives of increasing capacity, environmental flows in Summer and security of supply, the local authority: Mulwarree Shire Council decided to:

- Construct a 35 ML reservoir to ensure water availability during dry periods.
- Negotiate with a private landholder to supply 300 ML/year to meet requirements from the Healthy Rivers Commission to meet environmental flows.
- Install a remotely operated microfiltration unit (1.5 ML/day) which would meet current and future water quality guidelines, remove waterborne parasites, cope with occasional algal blooms and which could be expanded to 2 ML/day.

5.2   Process Description and Performance

The quantum change with CMF-S is in the arrangement of the system so as to eliminate the need for pressure housings. The membrane modules are suspended in an open tank from a series of removable manifolds. On the Marulan plant there are a total of 56 modules installed on 7 manifold racks, each containing 8 membrane modules, as shown in Figure 2.

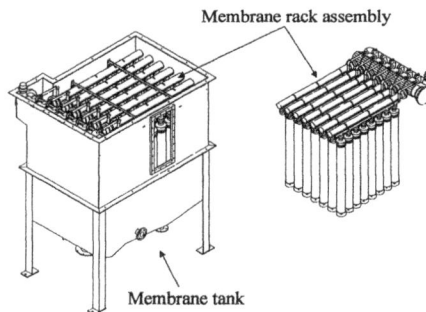

Figure 2. *Membrane tank at Marulan showing submerged membrane and membrane rack assembly*

5.3   Performance Result

The plant was commissioned in August 1998 and at the time of writing has been operating for just over 12 months. Demand on the plant has been relatively low compared to the design flow of 1.5 ML/day and the unit is usually only required to operate for 4 – 8 hours a day. The low demand and reduced flux means that the fouling rate is very low. The TMP stayed relatively stable for the first 6 months of operation, then an increase in demand saw the fouling rate increase. The first chemical clean was after 8 months of operation. Since then, the unit has been cleaned every 2 – 3 months, without the plant reaching high TMP, purely to ensure no regrowth in pipework.

Marulan 56G13 Performance

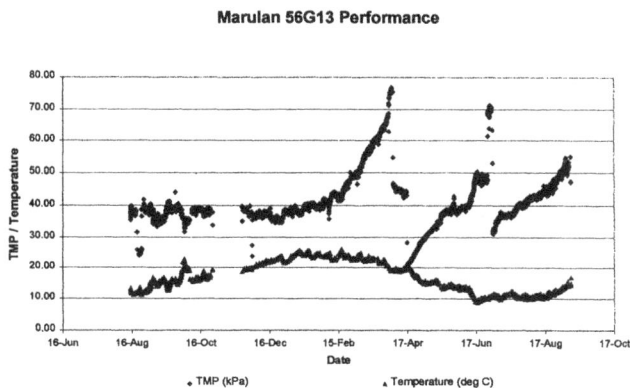

Figure 3. *Filtration Performance of CMF-S plant at Marulan WTP*

Operator involvement has been minimal. The plant is remotely monitored using a SCADA package and data is collected by an in-built proprietary datalogging system supplied with the CMF-S unit. The operator makes a daily visit to check the plant operation and chlorination system.

Pressure decay tests carried out on the unit indicate that the log removal of Cryptosporidium and Giardia will be at least 5 log. Regular pressure decay tests and an alarm if values exceed a preset limit, are used to ensure removal is always greater than at least 4 log.

6 OTHER APPLICATIONS

CMF-S has now been trialled on over 14 different sites with widely varying water and wastewater qualities. In all cases CMF-S has operated successfully, producing treated water of quality equal to conventional CMF in similar applications. As well as removing turbidity and organisms, coagulant has been directly dosed onto the CMF-S to remove colour and organic contaminants.

7 SCALE-UP OF CMF-S TO LARGE SCALE FILTRATION – AQUA 2000

7.1 Project Background

The first large scale CMF-S plant will be part of the AQUA 2000 project for Coliban Water, a 25 year BOOT scheme to provide drinking water to the Bendigo area in Victoria, Australia. USF Australia successfully tendered for this project with the CMF-S process in 1998. Currently design and construction is underway and commissioning is due late 2000.

AQUA 2000 project includes four water treatment plants spread over an area of 1600 square km, with a total capacity of 165ML/day. The largest plant, at Sandhurst Reservoir, will be the largest microfiltration plant in the world with an initial capacity of 126ML/day.

7.2   Process Description

The requirements of the BOOT contract have ensured the AQUA 2000 water treatment plants can cope with both current and future water quality requirements, most significantly in the area of removal of particles, algae and disinfection by-products.

The pre-treatment for the microfiltration uses Trident adsorption clarifiers, with coagulant dosing to remove metals and colour. Powdered activated carbon can be dosed in the event of algal blooms.

The building block of the CMF-S process is the 576 module cell, described in section 7.3. The largest plant in the AQUA 2000 project is at Sandhurst Reservoir and requires 6 x 576 module cells to reach the capacity of 126 ML/day.

USF Memcor Research tested the CMF-S and Trident adsorption clarifier processes on site at Sandhurst reservoir for eight months. The trial validated the combination of the combined Trident-CMF-S process, removal of particles, colour and organics – particularly the removal of disinfection by-products.

7.3   Large Scale CMF-S

The large-scale version of CMF-S is illustrated in Figures 4 to 6. The hollow fibre filtration membrane is fabricated into repairable **sub modules** having a nominal filtration area of 13 m^2. Four sub modules are arranged into a **sub manifold** assembly and up to eight sub manifolds connected via a stainless steel manifold to a removable **module rack**. The filtrate connection is via the top of each sub-manifold and backwash airflow via the bottom.

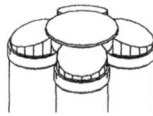

Figure 5 - 32 module rack

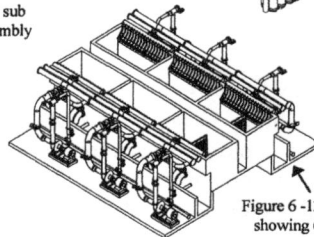

Figure 4 - Four sub modules in a sub manifold assembly

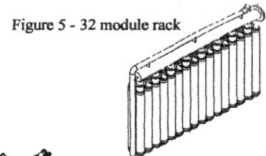

Figure 6 -120 ML/day plant showing 6 x 20 ML/day cells

Module racks are assembled into an open **filtration cell**. One back washable filtration cell can contain up to 18 module racks, i.e. 576 sub modules. Each cell has a nominal capacity of 20 ML/day when filtering typical surface water. Figure 7 shows the 6 x 576 module cells for the 126ML/day Sandhurst Reservoir plant.

The module racks are suspended in the cell and are lifted directly from the cell for servicing. For plant illustrated in Figure 6, two rows of three cells are configured back-to-back with a central feed channel in-between. Backwash channels are located at the opposite sides. Each cell incorporates a local filtrate pump and valves to permit individual flow control, backwash and CIP.

The CMF-S process has a small footprint. Each cell is approximately 6m x 3m. The 6 cell layout for the 126ML/day plant at Sandhurst Reservoir shown in Figure 7 occupies a space of less than 20m x 20m, excluding system ancillaries.

Filtration cell size and construction is adaptable based on a modular approach to meet specific application requirements. Individual cell size typically ranges from 160 to 576 total membrane sub-modules. The back-to-back configuration shown can be expanded to a 5 x 2 train that would have a nominal capacity of 200 ML/day. Multiples of this train are then used as building blocks for larger capacity systems. Total CMF-S plant capacity is not limited.

Trials are already underway to validate the 32-module rack design shown in Figure 5. These confirm exact scale-up from a single module to 32 modules (Figure 7). Scale-up from 1 rack to 18 racks (a full cell) is expected to be equally efficient. This is because:

The open tank design with submerged modules allows membranes to be in intimate contact with the feedwater at all times, regardless of cell size. Raw water distribution throughout the unit is assured by level.

Backwash air is distributed through manifolds directly into groups of four sub-modules so close control of air flow to each module is assured.

Solids backwashed from the membrane are completely removed from the system by draining the cell at the end of the backwash cycle. This eliminates any possibility of maldistribution within the cell that may lead to loss of scale-up efficiency.

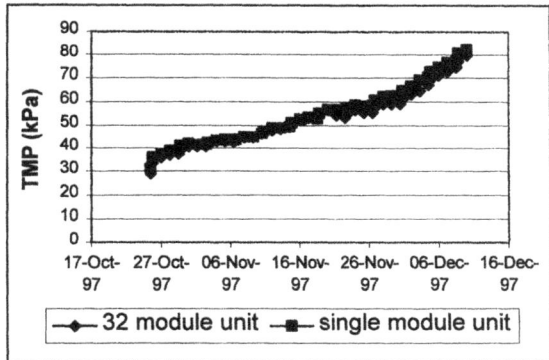

Figure 7 – *Scale-up from a single to module to a rack of 32 modules*

8   COSTS

As CMF-S is a submerged process, pressure vessels are not required to house the modules. This greatly simplifies the system design and reduces the capital cost. In addition the design of CMF-S lends itself to scale-up more readily as large membrane area can be installed in each cell.

The disadvantage of a submerged process is that it is TMP limited. It is not possible to run the system to TMP's greater than the pump design will allow, without adding liquid height above the membranes. The practical outcome is that the operating TMP range of CMF-S is less than in a pressurised process.

This means that a lower flux (or greater membrane area) is needed if the CIP interval is to be the same for both processes. A design flux for CMF-S between 75% and 95% of CMF is typical, depending on the feed water. However, any increase in membrane area is more than offset by a lower power cost (due to lower TMP) for the CMF-S process.

The power consumption for the CMF-S process is up to 50% of that for conventional CMF.

Chemical consumption is approximately the same, and maintenance costs are expected to be lower due to fewer valves and a less demanding operating cycle.

The net result is that the operating cost of the CMF-S process is usually less than for conventional CMF (typically about 5% lower).

9   CONCLUSIONS

- The CMF-S (submerged) series is the fourth generation CMF system and represents a natural progression of the CMF technology into very large-scale filtration applications. The simplification of the process and system allows it to be applied cost effectively at capacities greater than 100 ML/day using unit multiples of approximately 20 ML/day.
- The CMF-S process is to be used in the largest microfiltration plant in the world (126ML/day) as part of the AQUA 2000 project due to be commissioned late 2000.
- Many of the features of the conventional CMF process have been retained, including the use of air for scrubbing the membranes, integrity testing using pressure decay, and the same 0.2um polypropylene membrane, thus building on the experience of over 700 CMF installations worldwide.
- The Marulan plant represents the first commercial installation of the new fourth generation CMF-S process. Although rated at 1.5 ML/day the plant demand is a fraction of design flow. The plant has been running for an average of 6-8 hours per day at reduced flux since August 1998.
- CMF-S has now been trialled on over 14 different sites with widely varying water and wastewater qualities. Trials have confirmed that both the backwash efficiency and treated water quality for CMF-S are equal to conventional CMF in similar applications.
- Operating costs are overall lower than conventional CMF due principally to a significantly reduced power requirement. This is partially offset by a lower flux (higher membrane area per unit flow) due to the limited TMP range.
- The simplicity of the process, particularly for large scale systems, significantly reduces the capital cost of the system. Designs have been developed and component manufacture is underway for systems that use cells, with a nominal capacity of 20 ML/day, to build systems of greater than 100 ML/day cost effectively.

References

Hong, S.K.; Taylor, J.S.; Miller, F.; Rose, J.; Gibbson, C.; Owen, C.; Johnson, W.T. "Removal of Microorganisms by MF Process: Correlation between Integrity Test Results and Microbial Removal Efficiency". Proceeding of AWWA Membrane Technology Conference, Longbeach, CA, April 1999.

Water Quality and Treatment

THE UK SYSTEM OF APPROVAL OF PRODUCTS USED IN CONTACT WITH DRINKING WATER

Dr Toks Ogunbiyi

Technical Secretary
Committee on Chemicals and Materials
Drinking Water Inspectorate
Ashdown House
123 Victoria Street
LONDON SW1E 6DE

PREAMBLE

The Secretary of State has powers to make regulations controlling substances, products and processes used in the treatment and provision of public water supplies in England and Wales.

1 INTRODUCTION

The development of the regulatory framework for the authorisation/approval of products used in contact with public water supplies may be examined through three time periods – pre 1989, current/post 1989 and the future.

2 THE PERIOD BEFORE 1989

Prior to 1989, products used by water undertakers were introduced into the public water supply under a 'voluntary' system. The Committee on Chemicals and Materials did not have any statutory powers but was accepted as watchdog whose opinions were respected though not legally binding; it had bark but lacked bite.

The Committee was established in 1966 by the then Minister of Housing and Local Government. Prior to the Water Act 1989 the Committee operated on a non-statutory advisory basis, initially to approve chemicals for use in public water supplies. The scope of the Committee's activities was later extended to swimming pools, desalination plants and materials of construction for use in public water supplies. The Committee's terms of reference were further extended to consider the environmental impact of chemicals and materials submitted for approval. The Committee approved products if use in contact with water was considered **to be unobjectionable on health grounds** and where appropriate, the conditions of approval took account of the environmental impact of the product. The Committee periodically published statements listing approved products, the last being the 15th Statement, issued in 1989. Water undertakers were not legally obliged to use approved products but did so by and large.

3 THE PERIOD SINCE 1989

The advent of formal regulation of products used in contact with public water supply may be seen as the Water Act 1989. Section 53 gave the Secretary of State powers to make regulations controlling substances, products and processes used in the treatment and provision of public water supplies. Regulation 25 of the Water Supply (Water Quality) Regulations 1989 came into force on 1 September 1989. Regulation 25 concerns the introduction of substances and products for use in the treatment and provision of public water supplies. Under **Regulation 25(1),** a water undertaker shall not, otherwise than for the purposes of testing and research, apply to or introduce into water which is to be supplied for drinking, washing, cooking or food production purposes, any substance or product unless it:

(a) has been approved by the Secretary of State for the   Environment and the Secretary of State for Wales   [regulation 25(1)(a)]; or
(b) is considered by the water undertaker to be unlikely to affect adversely the quality of the water [regulation 25(1)(b)]; or
(c) has been used by a water undertaker during the 12 months prior to 6 July 1989 [regulation 25(1)(c)]; or
(d) is listed in the 15th Statement or any supplement issued by the Committee on Chemicals and Materials of Construction for Use in Public Water Supply and Swimming Pools [regulation 25(1)(d)].

Under **Regulation 25(4)** the Secretary of State has **powers to prohibit** the use of any substance or product which water undertakers were previously authorised to use under regulations 25(1)(b), (c) or (d). The Secretary of State may also revoke or modify conditions of any approval given under regulation 25(1)(a). Unless it is in the interests of public health for prohibition, revocation or modification to take effect immediately, six months' notice must be given in writing to water undertakers and approval holders.

Regulation 25(8) requires the Secretary of State at least once in each year to issue a list of all substances and products for which approval has been granted, refused, modified, revoked or prohibited.

Regulation 26 (1) provides for the Secretary of State to require a water undertaker to make an application for approval of any process or to prohibit the use of any process.

Regulation 26(4) makes provision for the publication of a similar list for processes.

Under **Regulation 27** the Secretary of State may require the applicant to pay a charge which reflects administrative expenses of determining an application for approval.

Regulation 28 relates to offences resulting from any breach of regulations 25 and 26, or from the provision of false information in support of an application for approval.

3.1 Situations where Regulation 25 is Not Applicable

Regulation 25 applies to substances and products which come into contact with water in the treatment and distribution of public water supplies, including use in raw water storage reservoirs, borehole installations and pipelines. Use in the following circumstances falls outside the scope of regulation 25:
(a)   after the time of supply i.e. within consumers' installations;
(b)   in treatment and distribution of private water supplies;
(c)   in treatment of swimming pool water; and
(d)   application of pesticides and weed control agents to raw water storage areas.

Use of substances and products in circumstance (a) falls within the scope of the Water Regulations Advisory Scheme. Although there is no legal requirement to use approved substances and products in circumstance (b) the Inspectorate has advised that this be done wherever practicable.

The Committee continues to operate a non-statutory approval system for substances and products used in swimming pools. The voluntary approval scheme for chemicals used in the treatment of drinking water by membrane and distillation processes has been discontinued. All products approved under this category are to be reviewed for approval under Regulation 25(1)a.

Circumstance (d) falls within the scope of the Control of Pesticides Regulations 1986.

3.2 The Committee on Chemicals and Materials

The Committee on Chemicals and Materials was recently re-constituted in the light of the Nolan Committee's recommendations, as a non-departmental public body. It now consists of a Chairman and five members with a range of expertise including engineering, materials science, toxicology and water treatment; an independent member represents water consumers' interests.

Products are recommended for approval if they are considered to be unobjectionable on health grounds, in the light of the best available evidence at the time. Approval is given solely for the purposes of Regulation 25 of the Water Supply (Water Quality) Regulations and should not be taken to imply any recommendation as to the technical merits of product in contact with water for public supply.

Products such as reverse osmosis, nano-, ultra- and microfiltration membrane elements and systems are usually tested for their leaching characteristics.

The Committee carries out a risk assessment and where the nature and level of the leachate determinands are such that they are considered to be of low or negligible risk to the consumer such a product may be recommended for approval by the Secretaries of State for the Environment and for Wales.

The Drinking Water Inspectorate does not grant the approvals. Accordingly the use of the DWI logo for advertising or any other promotional material is not allowed.

3.3 Cost of Approvals

The Secretary of State **currently** does not charge for the approval of products. Applicants should however expect to pay the designated test laboratories for any tests the laboratories may carry out in respect of their products.

3.4 Scotland and Northern Ireland

The Water Supply (Water Quality)(Scotland) Regulations 1990 and the Water Quality (Northern Ireland) Regulations 1994 make provisions for the control of substances and products. These regulations do not contain provisions for separate approval systems, although the Secretary of State for Scotland issues duplicates of approval letters and other legal instruments issued in England and Wales. Approvals issued by the Secretary of State are recognised in Northern Ireland, but separate approval letters are not issued there.

4 FUTURE DEVELOPMENTS

The development of legislation in the UK has inevitably been influenced by developments in the EU. This is in part due to the requirements of the Single European Act and the need to remove technical barriers to trade; but also as a public health measure.

4.1 The Role of CEN Standards

European Standards for water treatment chemicals are being developed by Working Group 9 of CEN Technical Committee 164. These standards are published by the British Standards Institution in the BS:EN series. Subject to complying with national conditions of approval, chemicals for drinking water treatment which conform with a BS:EN may be used without the approval of the Secretary of State.

It is envisaged that in future, water utilities will be able to make greater use of European Standards when procuring chemicals and construction products. There will be a need to make to make an assessment of whether the European Standard and manufacturer's intended use of the product are appropriate. Water utilities will be able to use any product which they are satisfied meets the European Standard without recourse to the Committee on Chemicals and Materials. This should ultimately eliminate the need for the remaining traditional use exemptions.

4.2 The End of Traditional Use Exemptions

DWI Information Letter 9/97 confirmed that the intended prohibition of unapproved uses of substances and products in water supply pipes, service reservoirs and water towers will be effective from **1 April 2000.** So far relatively few manufacturers and suppliers have made applications for the approval of these products. The prohibition removes the traditional use exemption which exists for cementitious pipes and linings, and bituminous linings, etc. From 1 April, 2000 all such products will have to be approved under Regulation 25(1)a.

Those UK manufacturers and suppliers who seek and obtain approval for all such products will be advantageously placed for the advent of the European Approval System (EAS).

4.3 Introduction of the EAS

For the past ten years there have been efforts aimed at harmonising the approval systems across Europe, for construction products (pipes, storage tanks, reservoirs, etc) used in contact with public water supplies. DG III of the EC has recently announced an initiative which will ensure closer alignment between the approval systems of all member states resulting in the development of the EAS. The EAS will operate through national regulatory bodies resulting in a 'European common list' of mutually recognised products which will be eligible for use by all water utilities throughout Europe without any further testing or assessment. This 'common list' is likely to be supervised by a system of regular audit inspections for material sources and manufacturing procedures.

After the inception of the 'European Common List', UK applicants will still able to get their product onto that list by application to one of the national regulatory bodies which operates the EAS. In the UK it is likely that the Committee on Chemicals and Materials will fulfil the national regulatory role with its activities extended to cover products used within consumers' premises.

IMMERSED MEMBRANES FOR DRINKING WATER PRODUCTION

Pierre Côté

ZENON Environmental Inc.,
845 Harrington Court,
Burlington
L7N 3P3,
Ontario,
Canada

Christian Güngerich and Ulrich Mende

Zenon GmbH
Nikaulos-Otto- Str.4,
D-40721,
Hilden,
Germany

1 INTRODUCTION

In the passed ten years, membrane filtration has been proven as a technology offering high quality, safety and reliability for the production of drinking water. Over the same period, prices have decreased by a factor of more than five, making membrane filtration cost competitive with conventional clarification and disinfection technologies.

There have been intense and rapid technological changes over the same period, but membranes and systems have not been standardised yet. Some key choices have been made: membranes are in the range of loose ultrafiltration to microfiltration, with pore sizes ranging between 0.01 to 0.2 μm, materials are polymeric, and the configuration is the hollow fibre ranging in diameter from 0.5 to 2.0 mm. However, for about half of the products available, filtration is from the inside-out, while for the other half, it is from the outside-in. In addition, most membranes are housed in pressure vessels and the feed is pressurised; the alternative configuration is a shell-less module with the driving pressure applied by suction on the permeate side.

Zenon Environmental Inc. has pioneered the development of shell-less immersed membranes. These authors are convinced that this configuration is best technological platform for a wide variety of applications in water treatment. This paper examines the key product and operating features of ZeeWeed®, Zenon's immersed membrane, presents an analysis of the benefits to users and describes various full-scale applications.

2 IMMERSED MEMBRANE AND MODULE

The ZeeWeed® hollow fibre has a composite structure illustrated in Figure 1. A strong support provides strength and flexibility, while the membrane permeation properties can be optimised separately. The inside and outside diameters of the hollow fibre are 0.9 and 1.9 mm, respectively.

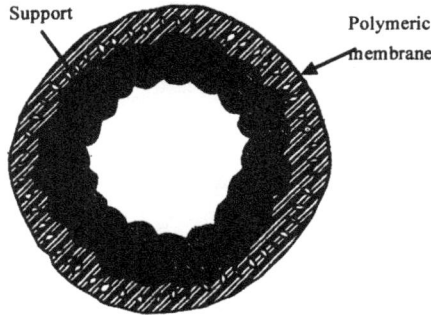

Figure 1   *The ZeeWeed® hollow fibre structure*

Various membranes are available for different applications. The OCP chemistry used in drinking water production, is an ultrafiltration membrane with a pore size of 0.04 μm. it is hydrophilic, chlorine-resistant and can be cleaned in a pH range of 2 to 10.5.

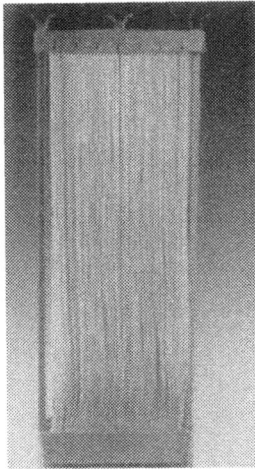

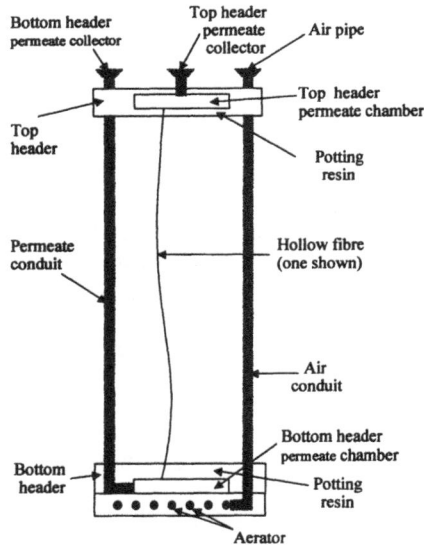

Figure 2   *The ZW-650 module*

A module consists of hollow fibres mounted on a vertical frame with permeate extraction from bottom and top headers (Figure 2). Each header contains a layer of potting resin that the fibres cross so that their lumens are connected to chambers where the permeate is collected. The hollow fibres are slightly longer than the distance between the top and bottom headers. Two rigid rectangular pipes connect the two headers and keep them at a fixed distance. One of them is used to carry the permeate from the bottom header to the top header. The other one is used to carry air to the aerators integrated into the bottom header.

The module dimensions are length of 2 000 mm, width of 700 mm and thickness of 200 mm. Two to twelve modules are assembled into a cassette, which constitute the building block for systems. A standard cassette of 8 modules contains 480 m^2 of membrane surface area and has a capacity of 30 to 40 m^3/h under a trans-membrane pressure up to 50 kPa.

3 IMMERSED MEMBRANE SYSTEM

An immersed membrane filtration plant is composed of parallel tanks into which cassettes are immersed. One tank normally corresponds to one production unit (one train) and is equipped with dedicated permeate pump and blower. This approach allows the construction of large trains of up to 20 000 m^3/d. When compared to pressurised membrane systems, where the train size is limited to approximately 4 000 m^3/d, this greatly limits the number of ancillary equipment, pumps, blowers and valves.

This design is very flexible and can easily be upgraded for flow by simply adding cassettes into the tanks. Where it does not freeze, the membrane tanks do not need to be covered by a building.

A complete system also includes a number of shared ancillary equipment: back-pulse pumping unit, permeate storage tank, clean-in-place unit and membrane integrity testing unit.

Feeding is normally by gravity and requires minimum pre-screening of 10 mm.

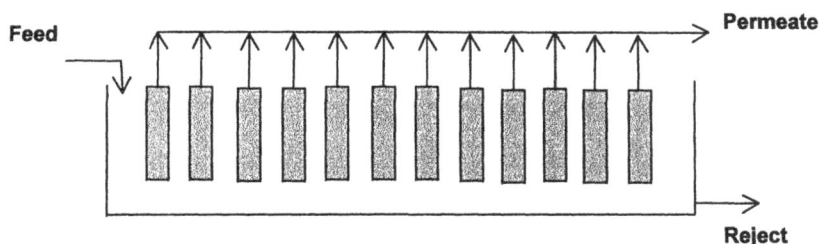

Figure 3   *Immersed membrane tank design with plug flow*

Small systems are operated as completely stirred tank reactors (CSTR). However, the common design for larger systems is to implement plug flow in a long tank (PFR), feeding at one end and continuously bleeding the purge at the other end, as illustrated in Figure 3.

This unique plug flow design offers significant benefits. First, it allows continuous withdrawal of concentrate without interruption of production. Secondly, high recovery (the fractions of feed recovered as permeate), up to 99% is easily obtained in a single step. Thirdly, the feed water gets concentrated as it flow along the tank and only the downstream cassettes see the most concentrated water. This is illustrated in Figure 4 where the CSTR and PFR configurations are compared for feed water containing 10 mg/L of suspended solids.

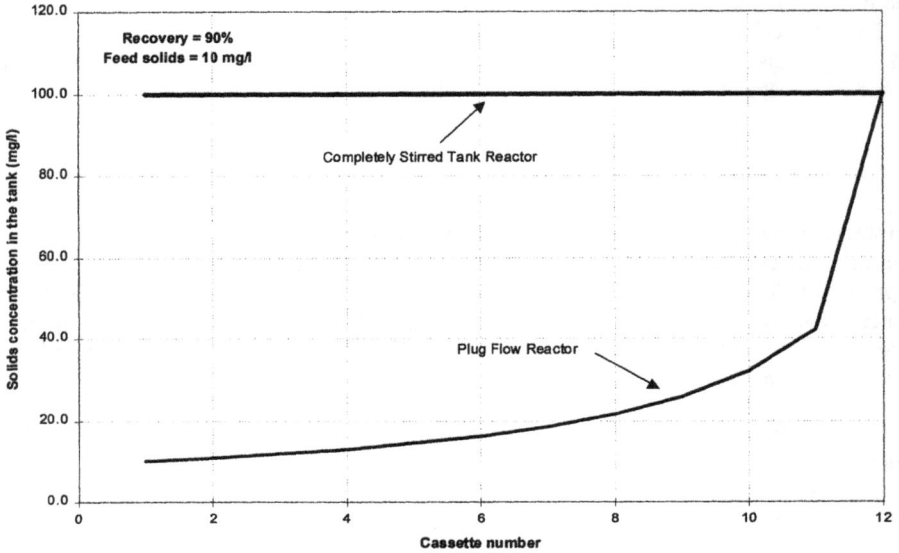

Figure 4   *Concentration gradient in the membrane tank for the CSTR and PFR modes of operation in a 12 cassette tank*

4 IMMERSED MEMBRANE OPERATION

The history of membrane filtration technology is a succession of improvements where higher fluxes have been sought to reduce cost. However, these efforts have been mostly directed at reducing capital cost of pressurised membrane systems where the membrane modules have low packing density and require a lot of ancillary equipment. When the cost of implementing membrane technology is examined with a life cycle analysis approach, it becomes clear that energy and membrane replacement costs are significant factors.

The design philosophy for a ZeeWeed® system is to reduce the operation requirements to the level of simplicity of conventional water treatment equipment such as sand filters. This is achieved by operating the membrane in quasi-direct filtration (dead-end), at low trans-membrane pressure and moderate fluxes. Gentle aeration agitates the hollow fibres and induces an airlift circulation pattern in the tank. Periodic back-pulses are used to de-concentrate the surface of the membrane.

This mode of operation corresponds to low fouling conditions. First, there is no pressure loss on the feed side of the membrane and the trans-membrane pressure is practically uniform along the length of each hollow fibre since permeate is withdrawn from both ends. Secondly, operation under suction imposes a practical limit to the trans-membrane pressure and operating flux, which remain in the stable pressure-controlled region of the filtration curve.

These gentle operating conditions correspond to low energy consumption (0.05 to 0.15 kWh/m³, depending on the application) and a long membrane life.

The mode of operation of a ZeeWeed® plant corresponds to low chemical cleaning requirements of 2 to 6 times per year. Cleaning is always done in-situ, using the back-pulsing unit to deliver the chemical solution to the membrane. The primary cleaner is chlorine, which is a common chemical in drinking water plants; the cleaning mixing water does not need to be heated. Cleaning can be done into an empty tank or a full tank. Empty tank cleaning offers the benefits of using full strength cleaning solutions and generating lower volume of waste. Full tank cleaning offers the possibility of prolonging the soaking time.

Membrane integrity is verified with an automated pressure decay test. This test is conducted by pressurising the lumen side of the membranes with air and measuring the decay in air pressure over time. The location of compromised fibres can be determined by observing the stream of air bubbles on the top of the membrane tank. Since the fibres are not in a shell, compromised fibres can easily be repaired in the field.

5 IMMERSED MEMBRANE APPLICATIONS

The ZeeWeed® membrane was first commercialised in 1994. The rapid development of the technology is summarised in Table 1, which shows the number of plants operating, and under construction or design. Initial plants were small (< 100 m³/d) focused on wastewater treatment using membrane bioreactors (i.e. the replacement of clarification by membrane filtration in the activated sludge process). ZeeWeed® started to be used for drinking water production in 1996. Currently, Zenon is designing a 100 000-m3/d ultrafiltration plant for Olivenhain, CA.

Table 1 *ZeeWeed® full-scale plants (Q3, 1999)*

Plant size (m³/d)	Number of Plants			
	< 100	100-1 000	1 000-10 000	> 10 000
Industrial				
Process Water	1	4	2	0
Wastewater	9	22	1	0
Municipal				
Drinking Water	1	9	11	6
Wastewater	22	9	21	1
Total	**33**	**44**	**35**	**7**

Because the ZeeWeed® membrane is resistant to most oxidants and tolerant to high concentration of suspended solids, it can be coupled with a conventional process such as oxidation, coagulation, adsorption, or biological treatment, for the removal of both dissolved and particulate contaminants. Even for a simple direct filtration plant, this flexibility can be used to modify the level of treatment in response to variable water quality by complementing membrane filtration within the same unit process. For example, powdered activated carbon can be added to treat seasonal occurrences of taste and odour or pesticides. The following applications have reached full-scale development:

- Direct filtration of surface water;
- Coagulation coupled with filtration for cold and low alkalinity coloured surface water;
- Oxidation coupled with filtration for iron and manganese removal;
- Concentration of conventional water treatment plant residuals;
- Pre-treatment of surface water to reverse osmosis;
- Filtration of tertiary effluents for water reuse;
- Activated sludge coupled with filtration for wastewater treatment.

Many papers that describe the performance of ZeeWeed® in various applications are listed in the reference section,[1,2,3,4,5,6,7,8].

6 CONCLUSIONS

Today, some ten years after commercialisation in the water field, membrane filtration is a proven technology. In essentially all new water treatment projects, membranes are now considered and are often selected over conventional technologies. However, the universal adoption of membrane technology by the water treatment community may be impeded by the lack of standardisation of configurations and operation modes. These authors submit that immersing membranes in a tank represents the best technological platform for standardisation, for the following reasons:

- Filtration with hollow fibres from the outside-in is more tolerant to variable feed water quality.
- Immersed membranes are conducive to the construction of large modules, cassettes and treatment trains, which significantly reduces the requirements for ancillary equipment, and thus translate in lower costs.
- Gentle filtration conditions, at low trans-membrane pressure and limited flux lead to low fouling rates and ease of operation.
- Immersed membranes are flexible tools that can be used for direct filtration or coupled with a conventional process such as oxidation, coagulation, or adsorption for the removal of both dissolved and particulate contaminants.
- Immersed membranes can be used to build new plants or to upgrade existing plants by immersing the membranes directly in clarifiers or sand filters.

References

1. S. Kroll, B. Vestby, *The Production of Ultra High Purity BFW from a Variable Surface Water Supply using Membrane-Based Technology*, presented at Power Gen International, New Orleans, December 1, 1999.
2. G. Best, D. Mourato, M. Firman, S. Basu, *Application of Immersed Ultrafiltration Membranes for Colour and TOC Removal*, presented at the AWWA Annual Conference, Chicago, June 22, 1999
3. H. Husain, P. Côté, *ZenoGem: The Zenon Experience with Membrane Bioreactors for Municipal Wastewater Treatment*, presented at the 2nd International Meeting on Membrane Bioreactors for Wastewater Treatment – MBR 2, Cranfield University, June 2, 1999.
4. P. Côté, D. Mourato, C. Güngerich, J. Russel, E. Houghton, *Desalination*, 1998, **117**, 181.
5. T. Lebeau, C. Lelièvre, H. Buisson, D. Cléret, L. Van de Venter, P. Côté, *Desalination*, 1998, **117**, 219
6. P. Côté, S. Monti, L. Belli, D. Bonelli, *The ZenoGem Process for Pharmaceutical Wastewater Treatment*, presented at the IAWQ 2nd International Conference – Advanced Wastewater Treatment, Recycling & Reuse, Milan, September 14-16, 1998.
7. Mourato, D., Benson, M., Carscadden, G., *The Role of Particle Count Analyzers as an On-line Integrity Tool for Immersed Membrane Plants*, presented at the AWWA Annual Conference & Exhibition, Dallas, June, 21-25 1998.
8. P. Côté, H. Buisson, C. Pound, G. Arakaki, *Desalination*, 1997, **113**, 189.

PHOSPHATE AND IRON REMOVAL FROM SEEPAGE AND SURFACE WATER BY MICROFILTRATION

J.A.M.H. Hofman[1], N.C. Wortel[2], E.T. Baars[1],J.P. van der Hoek[1]

[1] Amsterdam Water Supply
Provincialeweg 21
1108 AA AMSTERDAM
Phone: +31 20 6510200
Fax: +31 20 6976880

[2] Grontmij Consulting Engineers
P.O. Box 14
3730 AA DE BILT
Phone:+ 31 30 6943210
Fax: +31 30 6956366

1 INTRODUCTION

Drinking water for the City of Amsterdam and its surrounding municipalities is produced at two production plants. The total capacity of both plants together is 101 million m^3/y. The production plant East, also called the River-Lake Water Works, uses mainly seepage water from the Bethune polder (BP)(25 Mm^3/y). In dry seasons an additional in-take from the Amsterdam-Rhine Canal (ARC) is used (approx. 6 Mm^3/y). The seepage water from the Bethune polder originates from area with peat-soils, resulting in a relatively high organic content, a high colour, and presence of ammonia, iron and phosphate.

The water is pre-treated at treatment plant 'Loenderveen' with a double coagulation and sedimentation, storage in a lake, rapid filtration. The purpose of the coagulation system is mainly phosphate and turbidity removal. This is necessary to prevent eutrophication of the storage lake. A low phosphate level (20 ± 5 µg/l P) is however desired to maintain the self-purifying effect in the lake. The storage lake itself has three major objectives: removal of ammonium by nitrification, improvement of hygienic quality and damping of water quality peaks (mainly chloride from the ARC).

After pre-treatment the water is transported over a distance of 20 km to the Weesperkarspel treatment plant were the main treatment to drinking water quality takes place.

To fulfil future drinking water demands and improve drinking water quality, research is conducted to optimise the existing treatment process. Also treatment capacity extension is investigated for the River Lake Water Works. To increase capacity, more ARC water will be treated. When the existing treatment plants at Loenderveen and Weesperkarspel were designed in the 70's, it was expected that the treatment capacity would grow to a maximum capacity of 61 Mm^3/y. Parts of the treatment infrastructure were already built at this final capacity. However, growth of drinking water demand is not as rapid as was expected in the 70's. Therefore capacity extension in small flexible steps (around 2 Mm^3/y) may be beneficial. Because of its modular design, membrane filtration offers this flexibility.

For the pre-treatment system, microfiltration and ultrafiltration technologies offer good alternatives. Ultrafiltration was investigated earlier[1]. Results indicated good water quality as well as a relatively stable operation. However ultrafiltration showed a relatively high chemical demand (hydrochloric acid, hydrogen peroxide, sodium hypochlorite).

This paper presents the results of indicative study on the application of a Memcor Cross Flow Microfiltration system on Bethune polder water and ARC-water. Characteristic of the Memcor system is its outside-in filtration type and its air back wash.

The study was conducted in close co-operation with Grontmij Water & Waste Management. The objectives of this study were:

- removal of phosphate
- removal of pathogenic micro-organism
- removal of suspended solids
- fouling behaviour of the membrane modules.

2 EXPERIMENTAL

2.1 Pilot Plant

The pilot plant consisted of a small Memcor CMF system, containing three M10C membrane modules. Its total membrane surface area was 45 m^2. Air backwash was applied every 25 minutes. This interval was chosen based on experience with the membranes on a comparable source. No optimisation of the backwash interval was done.

For flux enhancement and phosphate removal, a ferric chloride dosing system was present in the feed system. Ferric chloride was added directly before the feed pump. The raw water from the ARC or the Bethune polder was prefiltered with a 200 µm microstrainer.

2.2 Experimental Programme

The experimental period was divided in two parts: In the first eight runs Bethune polder water was treated; in the second five runs ARC water was used. During each run the pilot plant was operated at a constant flux with an air back wash every 25 minutes. Each run continued until the maximum allowable Trans Membrane Pressure (TMP) was reached. After that, the membranes were cleaned thoroughly, using detergent cleaning agents. Also, the ferric dosage was varied, to investigate the necessity for phosphate removal and to investigate its influence on the membrane fouling rate. Moreover, the feed and permeate water were sampled to study the water quality.

The membrane fouling rate was determined by dividing the TMP increase by the run length in days. Table 1 and Table 2 give an overview of the experiments conducted during both experimental periods.

Table 1. *Experiments conducted on Bethune polder water*

Run	1	2	3	4.1	4.2	4.3	5	6.1	6.2	7.1	7.2	8
Flux	89	89	60	60	80	70	70	70	30	50	60	60
Iron	0	0	0,67	0,46	0,42	0,40	0	0,44	0,40	0,40	0,40	0,40
KPa/day	22,2	38,3	19,7	4,5	37,8	9,4	120	??	0	8,7	9,0	13,2

Table 2. *Experiments conducted on Amsterdam Rhine Canal water*

Run	1.1	1.2	1.3	1.4	2.1	2.2	2.3	3	4.1	4,2	5
Flux	60	70	80	70	70	70	70	90	80	60	80
Iron	3,9	4,0	3,9	3,7	2	1,1	0	0	2	2	2
KPa/day	7,3	8,6	15,3	7,3	15,3	7,4	16	91	28	1,5	19

3   RESULTS AND DISCUSSION

3.1   Process Stability

Figure 1 shows the raw TMP data on the Bethune polder water treatment; results on ARC-water are similar. Each run can be clearly identified by the "saw teeth" in the graph. It is observed that for each operating condition (flux, ferric dosage), different fouling rates exist. In general it is found that fouling rate increases with flux and decreases by addition of ferric chloride to the raw water. Furthermore, it can be seen that most lines are more or less straight filtration curves, indicating the occurrence of 'normal' cake filtration.

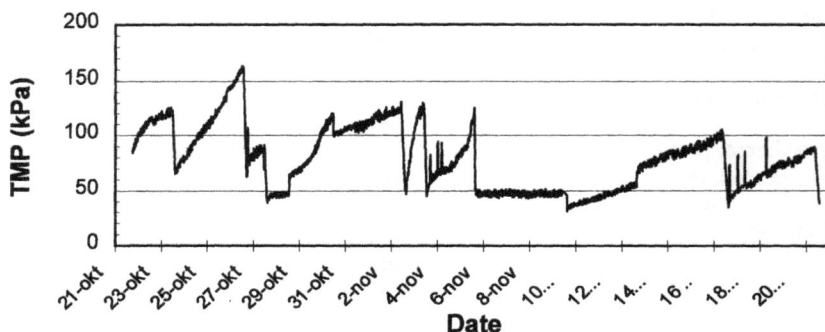

Figure 1. *TMP for the runs on Bethune polder water.*

Using these TMP data to calculate the membrane fouling rate and plotting them as function of the applied flux gives good opportunities for optimising the membrane operation. The results of this exercise are shown in Figure 2. It was found that the membrane fouling rate increases exponentially with the applied flux. Figure 2 shows that for ARC-water, the ferric addition reduces the fouling rate by a factor of approximately 2. For the Bethune polder water the effect of ferric addition is less clear, probably because of the natural iron content of the water.

The optimum flux can be found by selecting a chemical cleaning interval, e.g. 7 days. Using the maximum allowable TMP (180 kPa) and the clean membrane TMP (approx. 40 kPa), one can than calculate the fouling rate (20 kPa/d). Figure 2 than gives the flux, which can be applied: for the ARC a flux of 85 l/m^2h and 75 l/m^2h can be reached for operation with and without ferric chloride addition respectively. For the BP-water, the flux for stable operation is around 73 l/m^2h.

The results shown if Figure 2 indicate that a stable operation with the microfiltration can be achieved at relatively high fluxes and with low chemical demand. With ultrafiltration the same flux levels could be achieved, but with chemically enhanced backwashes every 3 hours and membrane disinfection every 24 hours [1] microfiltration however has higher energy consumption due to high pressure needed for the air back wash. Exact figures for the energy consumption however cannot be given on the basis of this study, because the further optimisation of the air backwash time interval is than necessary[2].

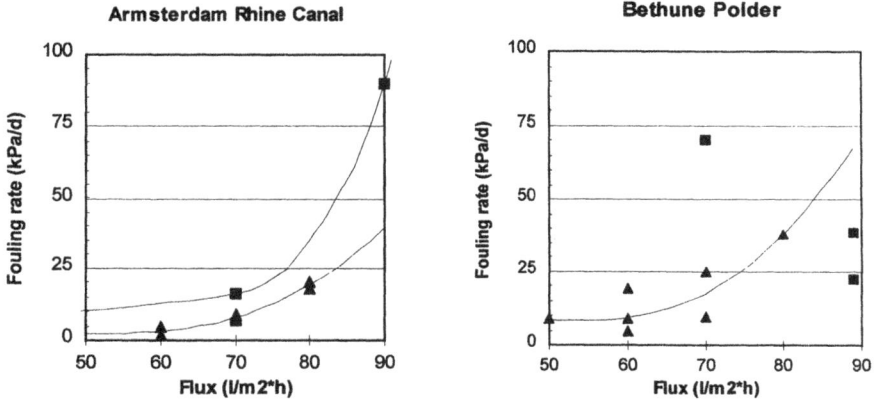

Figure 2. *Fouling rate as function of flux and ferric addition for both water types.* ■ *operation without ferric addition;* ▲ *operation with addition of ferric chloride.*

3.2   Water Quality

3.2.1   Treated water quality One of the major objectives of this study was to investigate whether microfiltration can be used for phosphate removal. Target value for phosphate is 20 ± 5 µg/l P. The levels in the raw water expressed as total-PO_4 are 100-200 µg/l P in the ARC-water and 50-100 µg/l P in the Bethune polder water. The phosphate in the ARC water is almost completely present as dissolved ortho-PO_4, This means that an ferric chloride addition is necessary to precipitate the phosphate in order to remove it by the membranes. The optimum ferric dosage was determined by varying the dosing rate and while on-line following the phosphate content of the microfiltrate. Figure 3 shows the results. From this graph it can be seen that a ferric dose of 2.5 mg/l Fe is sufficient to remove the o-PO_4 from the water. Further increase of the ferric dose will not affect the phosphate concentration in the microfiltrate.

For BP water, the phosphate is almost completely bound to the natural ferric in the seepage water. This means that no ferric addition is necessary for phosphate removal. Moreover ferric addition did not yield a flux improvement on BP water.

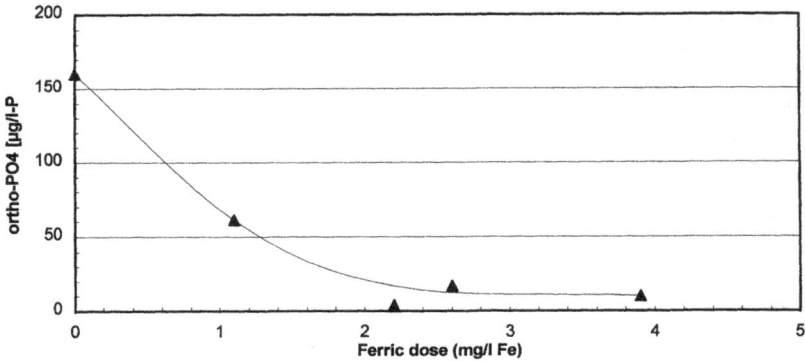

Figure 3. *Ortho-phosphate concentration in the ARC microfiltrate as a function of the ferric chloride dose.*

Beside the phosphate concentrations, other water quality parameters were determined. Average values are shown in Table 3. It is shown that microfiltration offers a very good water quality, as expected. Interesting point is that by the ferric addition, also DOC removal in the ARC-water was possible. In the BP water no removal of DOC was observed.

Table 3. *Overview of average water quality parameters*

Parameter	Unit	Amsterdam Rhine Canal		Bethune polder	
		Raw	Filtrate	Raw	Filtrate
Temperature	°C	8 – 6		10 - 6	
DOC	mg/l	5.0	4.5	10	10
Fe	mg/l	0.8	<0.1	3	<0.1
Mn	mg/l	0.18	0.14	0.2-0.4	0.1-0.3
As	μg/l	1	0.1	n.d.	n.d.
t-PO$_4$	μg/l	200-300	20-50	100-200	<20
o-PO$_4$	μg/l	140-160	20-40[2]	100-150	<10
Turbidity	FTU	12	<0.5	20	<0.5
SSRC[1]	N/100 ml	700	< 1	800-1000	<1

[1]   Spores of sulphite reducing clostridia
[2]   Depending of ferric dosage rate

3.2.2   Back wash water quality   Water quality of the back wash water and the Clean In Place (CIP) solution after cleaning was determined as well. For treatment of Bethune polder water iron, manganese and arsenic were important. Also DOC and AOX levels were measured because these parameters are required for disposal of the backwash and CIP solution. For the Amsterdam Rhine Canal, also heavy metals are important, since they are present in the raw water at significant levels.

Table 4. *Overview of average back wash water and CIP quality*

Parameter	Unit	Amsterdam Rhine Canal		Bethune polder	
		Back Wash	CIP	Back Wash	CIP
Turbidity	FTU	150	650	100	320
Suspended Solids	mg/l	65	65	160	220
Fe	mg/l	27	209	30	170
Mn	mg/l	0.44	0.84	0.33	1.02
As	µg/l	2	7	8	13
t-PO$_4$	µg/l P	0.1	195	5.6	215
o-PO$_4$	µg/l P	0.8	321	0.02	120.5
DOC	mg/l C	14.5	530	6.9	320
AOX	µg/l Cl	9	140	15	180
SSRC	N/100 ml			4800	16000

Table 5. *Heavy metals in back wash water and CIP of ARC*

Parameter	Unit	Amsterdam Rhine Canal	
		Back Wash	CIP
Aluminium	µg/l	590	18350
Barium	µg/l	180	560
Cadmium	µg/l	0.3	1.5
Chromium	µg/l	14	84
Copper	µg/l	19	171
Lead	µg/l	13	61
Nickel	µg/l	8	25
Vanadium	µg/l	15	39
Zinc	µg/l	70	460

Table 4 and Table 5 indicate that the back wash water contains a large amount of the filtered contaminants. However the CIP's, using a caustic surfactant solution, show that even more contaminants are released from the membrane surface. Because the CIP's were able to restore the original clean membrane flux it is assumed that they were effective.

4 INVESTMENT AND OPERATING COST

Based on the above results an estimation of investments and operation and maintenance costs was made for a extension of the River-Lake Water Works with 37 Mm^3/y. The total investment cost for this plant are estimated as 24.8 million Dfl (MEuro 11.3). This results in an O&M costs of Dfl 0.13 per m^3 (Euro 0.06 per m^3).

5 CONCLUSIONS

From the results it is concluded that microfiltration offers good possibilities for phosphate and turbidity removal from seepage and surface water in a cost effective way. Compared to ultrafiltration and conventional coagulation technology only limited amounts of chemicals are used, mainly due to a very effective air back wash system. A stable operation could be achieved during one month of experiments on each source. Addition of ferric chloride of 2-4 mg/l to the ARC water resulted in factor 2 reduction of the fouling rate of the membranes. For optimisation for a full-scale design more and long term research is necessary.

The results of this study will be used in a multi-criteria analysis for optimisation and extension of the River-Lake Water Works of Amsterdam Water Supply.

References

1. J.A.M.H. Hofman, M.M. Beumer, E.T. Baars, J.P. van der Hoek, H.M.M. Koppers, *Enhanced surface water treatment by ultrafiltration,* Desalination, 119 (1998) 113-125
2. J.A.M.H. Hofman, C.A. Groot, J.P. van der Hoek, *Ultrafiltration as alternative for coagulation and sedimentation for surface water treatment,* Proceedings of Water-symposium'99, Breda April 13th, 1999. (In Dutch)

REUSE OF FILTER BACKWASH WATER AS A SOURCE FOR DRINKING WATER PRODUCTION: PILOTING AND IMPLEMENTATION OF A FULL-SCALE ULTRAFILTRATION PLANT

A. Brügger, K. Vossenkaul, T. Melin, R. Rautenbach

Institut für Verfahrenstechnik, Rheinisch-Westfälisch Technische Hochschule Aachen
Turmstraße 46,
52062 Aachen
Germany

B. Golling, U. Jacobs, P. Ohlenforst

Stadtwerke Aachen AG,
Postfach 500155,
52085 Aachen
Germany

1 INTRODUCTION

The drinking water treatment plant Hitfeld of the Stadtwerke Aachen AG (STAWAG), Germany, processes 2.7 Mm³ reservoir water per year. After flocculation and pH adjustment the raw water passes a recirculation filter and a dual-media sand filter. Backwashing of these filters consumes about 10 % of the treated water. Since the processing capacity of the treatment plant exceeds the annual contingent of raw water from the dam, standstills of the plant may occur in dry years. In order to achieve a better plant utilisation and to avoid the costs of the wastewater, recycling of the filter backwash water is highly desirable.

Before recycling, the filter backwash water is stored in a basin for sedimentation. Until 1997 the clear water from the settling tank was then directly returned to the raw water. In compliance with a recommendation of the Drinking Water Commission of the Umweltbundesamt (Ministry for the Environment) the reuse of this water as raw water was stopped and the clear water was drained. Today the reuse of filter backwash water requires the separation of solids and moreover the complete removal of microorganisms and parasites.

The required secure retention of microbiological parameters cannot cost effectively be guaranteed by means of conventional technologies. The discussion of reusing the backwash water today often results in considering the use of membrane technology[1,2]. Therefore on behalf of STAWAG a pilot-scale ultrafiltration plant was operated by the Institut für Verfahrenstechnik (IVT) in co-operation with Rochem UF-Systeme GmbH, Hamburg (Rochem). The promising results of the pilot study led STAWAG to decide to treat the backwash water with a full-scale ultrafiltration plant. Here, results from the pilot study as well as first experiences with the full-scale plant will be presented.

2 PILOT-SCALE EVALUATION

2.1 Research Objectives

From the middle of June until the middle of August 1998, a pilot plant was set up in the drinking water treatment plant in Hitfeld in order to investigate the operational possibilities and the efficiency of ultrafiltration for the treatment of filter backwash water. The pilot plant was bypassed to the clear water disposal. Two full-scale modules treated corresponding to the incorporation of the future full-scale plant (Figure 1), the clear water.

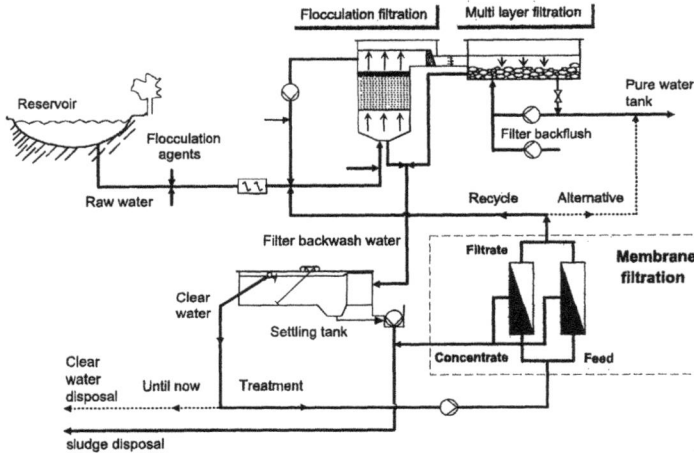

Figure 1 *Incorporation of the membrane plant into the drinking water treatment process*

Only working under real process conditions, including water quality and daily standstills caused by the necessary time for filter backwash and settling, give reliable results to the questions regarding the possibility of the treatment of the filter backwash water with the use of membrane technology. The aim of the pilot study was to provide results about the efficiency of a membrane plant, the optimisation of the plant operation and the establishment of design data for a full-scale plant. Therefore, the following questions had to be answered:

Which membrane material is most suitable for this application and can guarantee the production of the desired quality of filtrate in the long term? Which flux can be obtained on a long-term basis?

What are the optimum conditions for the operation of the plant and what recovery rate can be obtained?

How can the module flushing be fitted on to this application and how often are chemicals required for cleaning or disinfection?

How does the daily standstill of the plant influence its operation?

2.2 The Pilot Plant

Rochem developed the FM module system applied for the piloting in co-operation with the IVT. The principal elements of the FM module are membrane cushions, each consisting of two rectangular membranes; two internal permeate spacers, and an incorporated carrier plate, which are welded by an ultrasonic method on the outer edges (Figure 2). A membrane element consists of these membrane cushions, which are stacked one on top of the other supported by two pins with a gap in between. These cushions are enclosed in two halves of a shell. In addition to fixing the membrane stacks, the guide pins, in combination with corresponding bore holes in the half-shell elements, carry off the permeate[3].

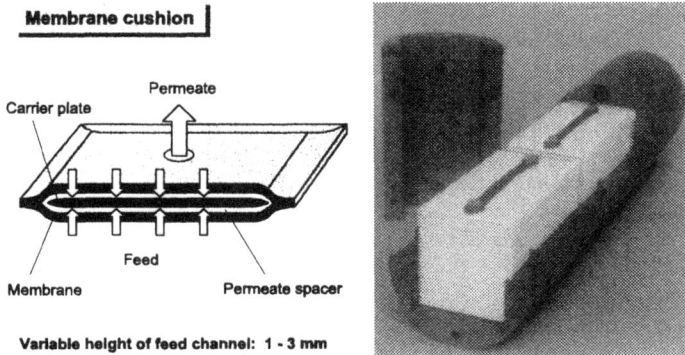

Figure 2 *Rochem FM Module System*

In the pilot plant eight of these membrane cushions elements are connected in a series and installed in a PMMA pressure tube. The pilot plant consists of two modules each fitted with about 7m² membrane area. Both modules can be operated separately and are installed vertically to allow for effective air flushing. In order to optimise the plant operation the pilot plant can be run in manual or automatic mode, where values for the main process parameters can be chosen separately for each module.

Since solid matter concentration in filter backwash water – especially after sedimentation – is quite low, dead-end mode was chosen for this application. This means that during filtration the water is flowed perpendicularly through the membrane. The module is run as a two-end-module. The retained solid matter forms a fouling layer on the membrane surface and, gradually, builds up a flow resistance, which adds to the flow resistance of the membrane. To avoid a decreasing permeate flux, the pressure difference has to be increased during the filtration cycle.

When the filtration cycle is complete the fouling layer has to be removed, i. e. the module has to be cleaned. The long-term stability of a dead-end filtration process relies on the efficiency of the cleaning procedure. The configuration of flat channels in the FM module allows a very efficient cleaning method consisting of a combination of feed sided air bubble flushing and back washing of permeate through the membrane. The air injected into the raw water at the base of the vertically installed modules induces high shear stresses and removes the fouling layer from the membrane surface. The subsequent short period of cross-flow flushing carries the detached particles out of the module.

2.3 Results of the Pilot Study

The pilot study was begun in early June 1998. At the beginning, both modules were equipped with polyaramid (PA) membranes. At the end of July polyacrylonitrile (PAN) membranes replaced the membranes in one of the modules. Both membrane materials had a cut-off of 50 kD.

The investment costs of dead-end membrane plants make up for the biggest part of the total specific treatment costs. Thus, increasing the plant capacity by enhancing the permeate flux means a substantial reduction for the total treatment costs. Therefore, the most important aim of the pilot study was to find out the highest possible flux at stable conditions with optimised process parameters. The parameters, which needed to be optimised, were the filtration time and all parameters related to the module flushing.

The start up of the pilot plant occurred at a relatively low flux performance. The flux could be raised by gradually adjusting the module flushing parameters to this application (Figure 3). Reasons for drops of the transmembrane pressure-difference (TMP) at constant fluxes were usually due to extra module flushes with air and water which were carried out in order to optimise the module flushing. As can be seen in Figure 3 it was possible to increase the performance of the PA membrane to over 150 l/m²h.

After membrane replacement the flux had to be reduced due to a rapid increase of TMP. Obviously the PAN membrane material can only be run at a somewhat lower level of flux. Applying a relatively high TMP a flux of about 115 l/m²h was possible at stable conditions with the PAN membrane.

During dead-end filtration the amount of produced filtrate equals the content of feed. Only the filtrate used for backwashing and the feed used for the cross-flow flush reduce the recovery rate. After a short start-up phase the plant reached a recovery rate of about 96 %. Recovery rate as well as energy consumption and the net permeate flux depends on the filtration time. At the end of the pilot study the filtration time for the PA membrane was 120 min and for the PAN membrane 100 min, altogether resulting in a high net flux and recovery rate at low energy consumption (Table 1).

It must be mentioned at this point, that during the three months of piloting, no chemicals were used for backwash enhancement or for chemical cleaning. The high fluxes were achieved only by adapting and optimising the module flushing procedure for this application.

The membrane material chosen to equip the full-scale plant was the PAN membrane. This was due to doubts as to whether the membrane cushions made from PA could guarantee long term integrity. Especially the welded edges of the PA membrane cushions were susceptible to damages. It was observed during the pilot study that the performance of the PA membranes regarding permeate quality was inferior to the PAN membranes. The quality of the permeate was determined by continuous particle counting and by plate counting. Although both membranes always showed drinking water quality and low particle counts there was a difference between the two materials. The retention of CFU for PAN membranes was about one log-stage better, and particle counts (>1 μm) for PAN were always below 0.5 particles per ml while the counts for the PA membrane were between 0.5 and 1.5 particles per ml.

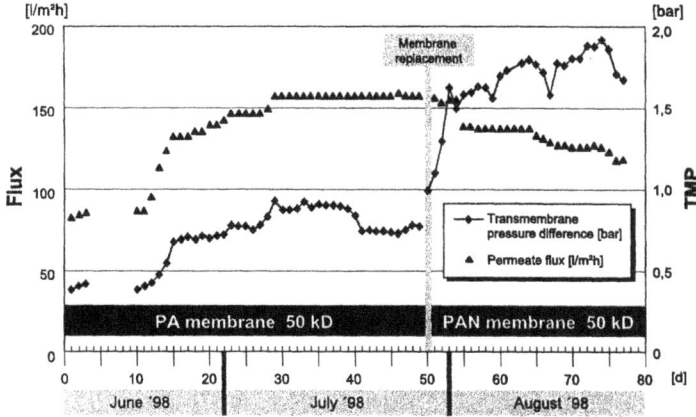

Figure 3 *Membrane performance in the pilot study*

Table 1 *Operating data of the pilot plant*

Operating parameters		Module 1	Module 2
Permeate flow rate filtration	[l/h]	1280	900
Membrane material	[-]	PA 50 kD	PAN 50 kD
Temperature of raw water (average)	[°C]	10	10
Duration of filtration cycle	[min]	120	100
Duration of flush	[min]	5,1	5,1
Membrane area per module	[m²]	6,7	7,2
Transmembrane pressure difference	[bar]	0,88	1,75
Permeate flux	[l/m²h]	190	119
Recovery	[%]	96,9	94,7
Net permeate flux	[l/m²h]	181	114
Energy consumption	[kWh/m³]	0,15	0,18

3 DESIGN AND COST CALCULATION FOR THE FULL-SCALE PLANT

3.1 Design Data

As a consequence of the positive pilot results STAWAG decided to employ the membrane filtration technology to recycle the filter backwash water. It was agreed that STAWAG would buy, at a fixed price, the filtrate treated by the plant owned by Rochem UF-Systeme GmbH, and operated for further optimisation by the IVT. The plant capacity was fixed to be 10.000 m³ per month, i.e. about 350 m³ per day. The daily operating time was assumed to be about 18 hours.

At the time when the plant design took place, the manufacturer of the PAN membranes was offering the same material used in the pilot plant with higher cut-off values. Thus, an optimistic value for the permeate flux for the plant design seemed to be realistic and it was assumed that a flux of about 100 l/m²h could be achieved (Table 2). As a result the required membrane area is 200 m² installed into 20 modules.

Table 2 *Design data for the full-scale UF- plant*

Design data		
Daily treatment duty	[m³/d]	350
Operating hours per day	[h/d]	18
Capacity	[m³/h]	19,4
Membrane area per module	[m²]	10
Number of modules	[-]	20
Entire membrane area	[m²]	200
Expected life time	[a]	4
Permeate flux	[l/m²h]	97
Filtration cycle	[min]	90
Recovery	[%]	95
Energy consumption	[kWh/m³]	0,2
Operating pressure	[bar]	1,0

3.2 Cost Calculation

The calculation of the specific costs for the ultrafiltration treatment of filter backwash water is based on a time of depreciation of 10 years. The characteristic value of investment of 1900 DM per m² installed membrane area represents an average value of comparable plants offered by different manufacturers. The specific costs for chemicals and maintenance are based on empirical coefficients.

The total specific costs are about 0,6 DM/m³ (Table 3). These costs are relatively low for the treatment of filter backwash water[2], relying on realising the high assumed permeate flux. According to this calculation the depreciation contributes about 70% of the total costs. This is characteristic for membrane plants operated in dead-end mode.

4 FIRST EXPERIENCES WITH THE FULL-SCALE PLANT

In February 1999 the full-scale plant was set up and operation was started. The plant consists of two blocks with ten modules each. At start-up the plant was equipped with about 200 m² of the PAN 50 kD membrane. The permeability of this membrane for the first four month in operation is shown in Figure 4.

A breakdown of the permeate backwash pump after 24 days resulted in a sharp decline in membrane permeability. For one weekend the efficiency of the module flushing was negligible. Since one of the modules was equipped with a PMMA pressure tube it was possible to see thick fouling layers on the membranes which built up over the weekend. Attempts to recover the former membrane permeability by extra module flushing failed. Even further improvement of the module flushing by adapting the procedure to the conditions of the full-scale plant could not recover the membrane permeability to the former level.

Table 3 *Estimation of costs for the full-scale plant*

Plant - data		
Time of depreciation	[a]	10
Interest rate	[%]	7
Capital factor	[%]	14,24
Use of capacity	[%]	75
Capacity per year	[m³/a]	127000
Cost for membrane replacement	[DM/m²]	200
Electricity cost	[DM/kWh]	0,2
Investment cost		
Characteristic value of investment	[DM/m²]	1900
Investment cost	[DM]	380000
Specific cost		
Depreciation	[DM/m³]	0,426
Energy	[DM/m³]	0,040
Chemicals, maintenance and staff (all-inclusive)	[DM/m³]	0,065
Replacement of membranes	[DM/m³]	0,079
Specific overall cost	[DM/m³]	0,610

Thus, after about 3 month in operation a chemical cleaning was carried out. Preceding tests at the IVT led to the decision to use an alkaline cleaner and NaOCl as cleaning agents. Permeability after chemical cleaning was almost equal to the start-up value. Applying the improved module flushing it was possible to achieve a stable permeability of 50 l/m²h bar for the PAN 50 kD membrane.

In parallel to the full-scale plant the pilot plant was operated with a new PAN membrane with a cut-off of 200 kD. This membrane not only showed a very high permeability but also the same reliable retention for particles (>1 μm) and CFU as the 50 kD membrane. Therefore, in one block of the full-scale plant the 50 kD membranes were replaced with 200 kD membranes. This membrane was operated at a flux of about 105 l/m²h with a permeability of nearly 80 l/m²h bar. With this membrane it is now possible to achieve the plant productivity as required (Table 4).

5 SUMMARY

Secure retention of microbiological parameters and high fluxes, which allow economical plant operation, are essential for successfully applying ultrafiltration for recycling filter backwash water. High fluxes on a long-term basis in this case were achieved by improving and adapting the module flushing procedure to this application. As a result, the use of any chemicals is restricted to a chemical cleaning every 3-4 month.

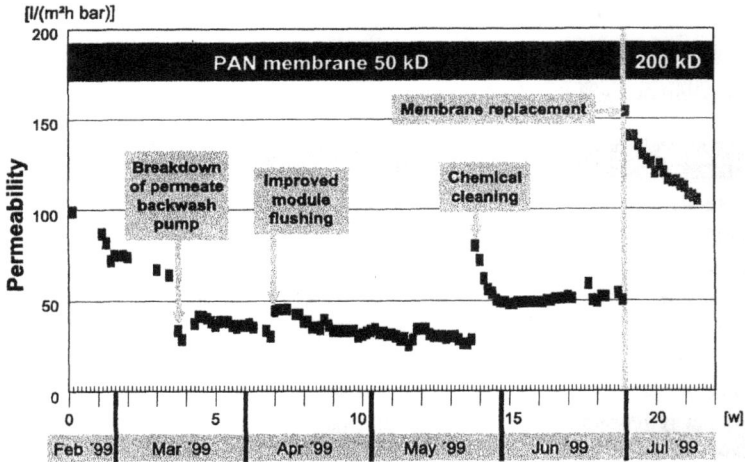

Figure 4 *Membrane permeability in the full-scale plant*

Table 4 *Development of productivity of the full-scale plant*

Date	Remarks	Flowrate [m³/d]	Daily productivity [m³/d]
18.02.99	Start-up	12	180
26.02.99	Increased flux	13	195
03.03.99	Increased flux	14	210
14.03.99	Breakdown permeate backwash pump	10,5	158
17.03.99	Reactivation of permeate backwash pump	13	195
06.04.99	Improved module flushing	14	210
24.05.99	Chemical cleaning	16	240
08.06.99	Improved module flushing	17	255
13.07.99	New 200 kD membranes in one block	19	285
15.07.99	Reduction of settling time	19	334

References

1. P. Lipp et al., *Desalination* 119 (1998) 133-142
2. R. Rautenbach, K. Voßenkaul, Proceedings *"2. Aachener Tagung Siedlungswasserwirtschaft und Verfahrenstechnik: Membrantechnik in der öffentlichen Wasseraufbereitung und Abwasserbehandlung"*, Aachen, 1998, A18
3. T. Peters, *Umwelt Technologie Aktuell International Edition*, 4 (1998)

IMPROVED PERFORMANCE OF DRINKING WATER MICROFILTRATION WITH HYBRID PARTICLE PRE-TREATMENT

T. Carroll and N. Booker

CSIRO Molecular Science,
Private Bag 10,
Clayton South MDC, VIC3169,
Australia.

1 INTRODUCTION

Microfiltration is particularly effective for the removal of particulate contaminants, such as bacteria, algae and protozoa from drinking water. However microfiltration is ineffective for the removal of dissolved contaminants such as natural organic matter (NOM). Residual NOM in drinking water is a cause of colour, disinfection by-products, and microbial regrowth[1,2]. Residual NOM in microfiltration processes is a cause of membrane fouling[3]. The conventional approach to removing NOM from drinking water is chemical coagulation and flocculation. This approach is now routinely used in Australia as a microfiltration pre-treatment to enhance colour removal and to reduce the rate of membrane fouling. However, there is considerable scope for further improvement by combining chemical coagulation with adsorption of NOM onto a solid particle. The adsorbent could be selected on the basis of low hydraulic resistance to filtration, and could be used under conditions in which it acts as a site for flocculation. The hybrid adsorbent-coagulant particle formed in this way would then be removed by microfiltration in the conventional way, depositing a more permeable filter cake on the membrane and producing a higher quality drinking water than coagulation alone.

Magnetite is a commonly mined mineral in Australia with potential as an adsorbent in a hybrid microfiltration pre-treatment. Magnetite is a magnetisable iron oxide used alone or in conjunction with alum or polyelectrolyte in an existing treatment process to remove particulate and some dissolved contaminants from drinking water[4]. It has a low surface area ($2\text{-}3m^2/g$) and therefore requires low contact times (< 60s) but relatively high doses (> $1g/L$). In addition, particles are heavy (s.g. = 5.1), and can be magnetically flocculated to settle rapidly (> 30m/hr). The adsorbent is easily regenerated by a caustic wash. These properties would facilitate in-line adsorbent dosing in an existing microfiltration plant. Removal of the flocculated adsorbent is simple and rapid, reduces any additional particle loading to the membrane, and aids recovery for reuse. The potential of magnetite as an adsorbent particle for use in combination with alum and polyelectrolyte as a microfiltration pre-treatment for Australian surface water is investigated in this paper.

2 METHODS

The raw water source was the Moorabool River near Anakie, Australia. This water has a relatively high total organic carbon content (8-10mg/L) and relatively low turbidity (3.9 NTU). The water was treated in standard jar tests with various combinations of magnetite (Fe_3O_4), alum ($Al_2(SO_4)_3.18H_2O$), and LT-20 polyelectrolyte (a high molecular weight, non-ionic polyacrylamide). The jar test procedure was to adjust the pH of 1.00kg of raw water from 7.5 to 6.0 with sulphuric acid. The water was then dosed with magnetite, alum, and polyelectrolyte as required. After each chemical addition the water was stirred for 60s at 250RPM. The pH was maintained at 6.0 throughout the procedure with sulphuric acid or sodium hydroxide. Once dosing was complete the mixture was stirred for 90s at 100RPM. The treated water was analysed for dissolved organic carbon (O/I Analytical 1010 Wet Oxidation TOC Analyser). All samples for water quality were filtered through a 0.45μm filter (Selby-Biolab HPLC-certified) prior to analysis. In hybrid magnetite-alum treatment cases, the extent of magnetite-alum attachment was measured from the residual aluminium content in the treated water by inductively-coupled plasma (Jobin-Yvon JY24 ICP spectrometer) after the jar contents were allowed to settle for 150s.

Microfiltration experiments were performed on the single-fibre filtration rig shown in Figure 1. The membranes were polypropylene hollow fibres with a nominal pore size of 0.2μm, an internal diameter of 250 m, an external diameter of 550μm, and a length of 1.2m. Fibres were sealed at the open ends with a silicone septum, wet with ethanol and flushed with pure water. The feed water was pressurised with a peristaltic pump, and forced through the hollow-fibre membrane to emerge as permeate from the open ends of the fibre. The permeate was collected in a vessel mounted on an analytical balance. A pressure transducer measured feed pressure. The signals from the analytical balance (WT) and the pressure transducer (PT) were processed to calculate the permeate flowrate as a function of permeate throughput. The feed water was treated using the standard jar test procedure described above, although the 100RPM stirring regime was omitted. The feed suspension was stirred at 250RPM and the pH was controlled at 6.0 throughout the membrane fouling experiments, which took 2-3 hours. In some cases, the feed suspension was allowed to settle, and only the supernatant was pumped onto the membrane.

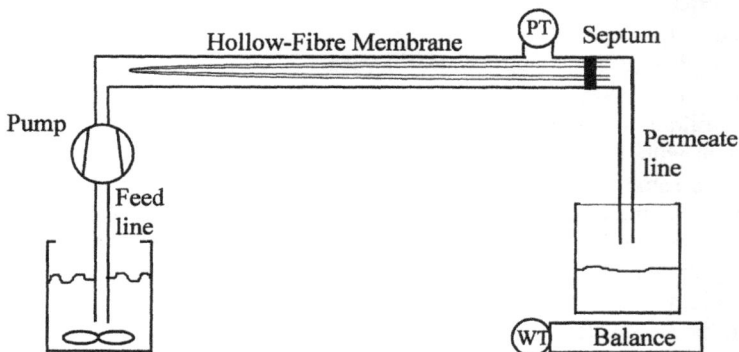

Figure 1 *Single-fibre microfiltration rig.*

3 RESULTS AND DISCUSSION

3.1 Magnetite-Alum Pre-treatment: NOM Removal and Membrane Performance

The dissolved organic carbon (DOC) concentration of Moorabool water treated with various doses of magnetite and alum is shown in Figure 2. The optimum alum dose (at 0g/L magnetite) for NOM removal was approximately 3.2mg/L as Al^{3+}. The DOC removal at this dose was 44%, and further addition of alum did not increase DOC removal. The combined magnetite-alum pre-treatment improved DOC removal over alum alone at alum doses below the optimum 3.2mg/L Al and magnetite doses of 1g/L or higher.

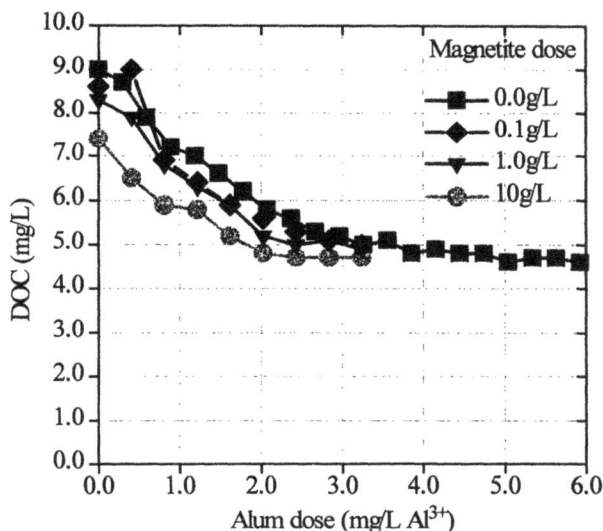

Figure 2 *Dissolved organic carbon of Moorabool water as a function of alum and magnetite dose.*

When magnetite was added before alum at pH 6, the alum flocs formed on the magnetite particles. This attachment occurred at relatively low magnetite doses, as shown in Figure 3. At an alum dose of 3.2mg/L Al, a magnetite dose of 0.01g/L was sufficient to attach 34% of the alum flocs formed (as Al^{3+}). At the optimum alum dose, and magnetite doses of 0.1g/L and above, the concentration of unattached alum flocs dropped below the concentration of aluminium in the raw water. Magnetite can therefore serve as a simple but effective carrier particle for alum floc thus reducing fouling of the membrane.

The membrane filtration performance of the untreated, alum-pre-treated, and magnetite-alum pre-treated water are shown in Figure 4. Microfiltration alone removed 11% of the DOC but the rate of fouling was relatively high; 35g of water was filtered before the permeate flowrate declined by 25%. Alum pre-treatment (3.2mg/L Al) increased the DOC removal to 44% and reduced the rate of fouling; 110g of water was filtered before the permeate flowrate declined by 25%. The rate of fouling was not further reduced by magnetite-alum pre-treatment at magnetite doses below 0.1g/L. However substantial

reductions in the rates of fouling were achieved for magnetite doses above 1.0g/L. At a
magnetite dose of 1.0g/L, 270g of water was filtered before the permeate flowrate declined
by 25%. At a magnetite dose of 10g/L, 460g of water was filtered before a decline of 25%.

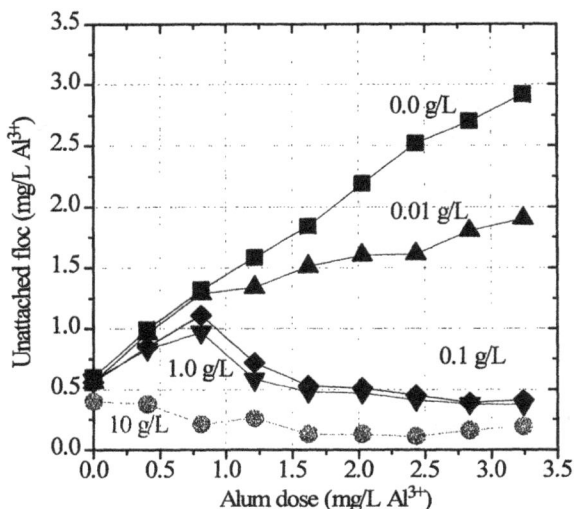

Figure 3 *Concentration of unattached alum flocs as a function of alum and magnetite
dose.*

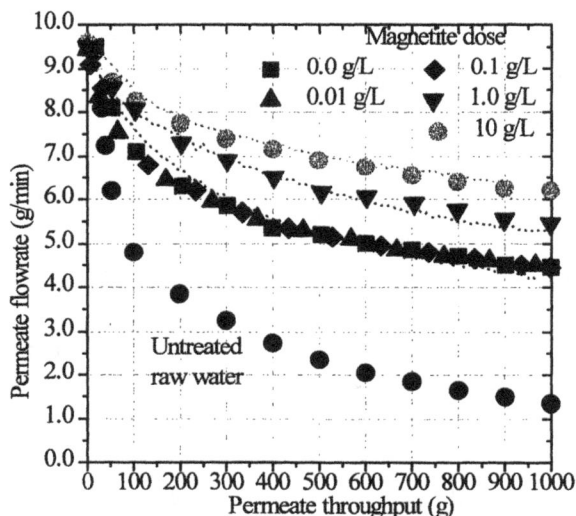

Figure 4 *The decline in permeate flowrate as a function of throughput for Moorabool
water treated with 3.2mg/L of alum (as Al^{3+}) and various doses of magnetite.*

Although the rate of membrane fouling was reduced by a combined magnetite-alum pre-
treatment, there was still a substantial decline in permeate flowrate with throughput during
microfiltration. Two possible causes of this decline are:

1. Residual NOM which is not removed by the hybrid pre-treatment subsequently fouls the membrane.
2. Particles and flocs generated in the course of treatment offer a hydraulic resistance to filtration when retained by the membrane.

To distinguish these two contributions, the alum-pre-treated and magnetite-alum pre-treated waters were allowed to settle before microfiltration and the supernatant was pumped onto the membrane. The membrane filtration performance of the settled pre-treatments are shown as dashed lines in Figure 4 for alum only, and alum with 1.0g/L and 10g/L of magnetite. There was no reduction in the rate of fouling in any of these cases relative to the corresponding stirred pre-treatments indicating that the decline in permeate flowrate must be due to suspended or dissolved material in the treated water. This material could be residual NOM or unattached alum pin flocs. Fouling by unattached alum flocs should be higher in stirred than in settled pre-treatment, and this was not observed. Furthermore, from Figure 3, the concentration of aluminium from which these flocs were constituted was relatively low (less than 0.5mg/L Al) after magnetite-alum pre-treatment and settling.

The reduction in the rate of membrane fouling achieved using a combined magnetite-alum pre-treatment is compared to alum pre-treatment alone in Figure 5. The basis for comparison is the permeate throughput corresponding to an arbitrary 25% decline in permeate flowrate (as an indicator of the throughput which could be achieved before intervention to recover performance would be required). On this basis, there was no reduction in the rate of fouling over alum for hybrid pre-treatments with doses of magnetite up to 0.1g/L. However at a magnetite dose of 1g/L, a 2.5-fold increase in throughput was achieved, and at a magnetite dose of 10g/L, a 4-fold increase in throughput was possible.

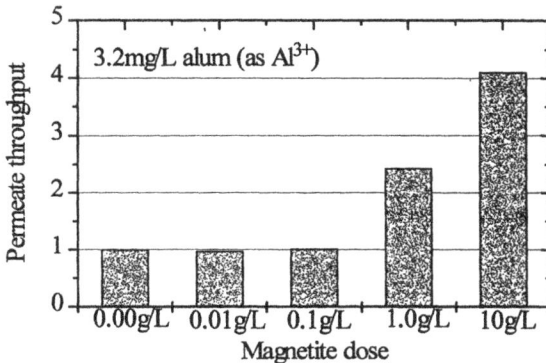

Figure 5 *The increase in relative permeate throughput for Moorabool water treated with 3.2mg/L of alum (as Al^{3+}) and various doses of magnetite.*

3.2 Magnetite-Alum-Polyelectrolyte Pre-treatment: Membrane Performance

The membrane performance was also investigated when a polyelectrolyte was added in addition to alum and magnetite. A polyelectrolyte may become necessary in full-scale processes to reduce magnetite-alum detachment and break-up under ubiquitous shear forces. The rate of fouling was unaffected at polyelectrolyte doses below 0.6mg/L but increased at higher doses. The reduction in the rate of fouling achieved using a combined magnetite-alum-polyelectrolyte pre-treatment is compared to alum pre-treatment in Figure 6. There is no improvement in the rate of fouling over alum alone for hybrid treatments with doses of magnetite up to 0.1g/L. However at a magnetite dose of 1g/L, a 3-fold increase in throughput was achieved, and at a magnetite dose of 10g/L, a 4.5-fold increase in throughput was possible.

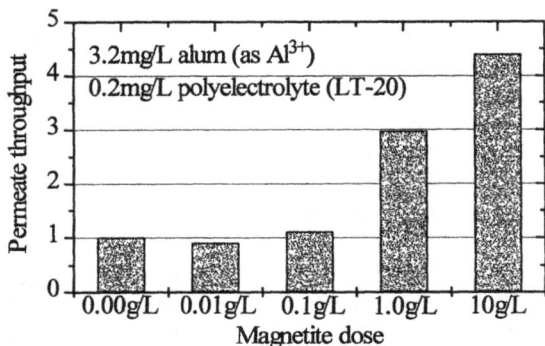

Figure 6 *The increase in relative permeate throughput for Moorabool water treated with 3.2mg/L of alum (as Al^{3+}), 0.2mg/L of polyelectrolyte, and various doses of magnetite.*

3.3 Magnetite-Alum-Polyelectrolyte Pre-treatment: Cost Benefit Analysis

The increased costs of the magnetite-alum pre-treatment are compared to the reduced cleaning costs for a microfiltration plant using only alum pre-treatment in Figure 7. The comparison was based upon the assumption that the benefit of higher throughput is reduced chemical cleaning frequency. Backwash frequency (and the associated downtime, reduction in product water yield, and energy consumption) was assumed to be unaffected. The capital and operating costs of the magnetite recovery plant were not considered. The cleaning costs are taken from a 6ML/day microfiltration plant operating with alum pre-treatment at Cresswell, near Melbourne, Australia. The additional treatment costs include sodium hydroxide for magnetite regeneration, polyelectrolyte for magnetite-alum flocculation, and magnetite to replace unrecoverable losses. From Figure 7 a doubling of microfiltration membrane throughput is required before the savings in cleaning costs balance the increased cost of the magnetite-alum treatment. From Figure 6, it can be seen that this throughput increase was achieved with a magnetite dose of 1.0 g/L at an alum dose of 3.2mg/L Al.

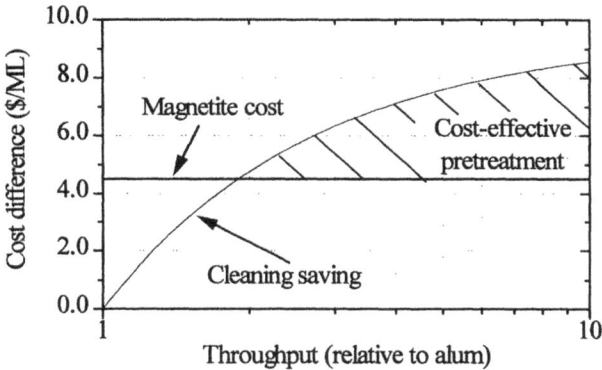

Figure 7 *The comparison between reduced cleaning costs and increased treatment costs associated with a hybrid alum-magnetite pre-treatment.*

4 CONCLUSIONS

A significant improvement in the laboratory-scale performance of drinking water microfiltration was achieved by combining a magnetite treatment step with conventional alum treatment at pH 6.0. The permeate throughput at which the permeate flowrate declined by an arbitrary value (25%) increased by up to a factor of four for magnetite doses up to 10g/L, although the water quality after magnetite-alum treatment was not substantially better than that achieved with alum alone. This improvement would translate into a four-fold increase in water production before the criterion for membrane cleaning or other remedial action is reached. The improvement in membrane filtration performance was independent of whether settling preceded filtration, or whether the treated water was filtered directly as a suspension. This may not be the case for an alternative water source, although settling was rapid, allowing a reduced particulate loading to the membrane if required. The rate of fouling after magnetite-alum pre-treatment was also unaffected by dosing with trace quantities of LT-20 polyelectrolyte, a flocculant aid which may be necessary to bolster the shear resistance of the magnetite-alum particles in a larger-scale treatment process. The hybrid alum-magnetite pre-treatment process is presently undergoing pilot trials on a 40kL/day microfiltration pilot plant at Yerring Gorge, near Melbourne, Australia.

References

1. J.J. Rook, Wat. Treatment Exam., 1974, 23, 234.
2. D. van der Kooij, J. AWWA, 1992, 84(2), 57.
3. A.B. MacCormick in Modern Techniques in Water and Wastewater Treatment, eds. L.O. Kolarik and A.J. Priestly, CSIRO Publishing, Australia, 1995, p. 45.
4. N.A. Booker, L.O. Kolarik and R.B. Brooks in Modern Techniques in Water and Wastewater Treatment, eds. L.O. Kolarik and A.J. Priestly, CSIRO Publishing, Australia, 1995, p.25.

RIVER TRENT ON TAP - COMPARISON OF CONVENTIONAL AND MEMBRANE TREATMENT PROCESSES

B.E. Drage and J.E. Upton

Severn Trent Water Ltd,
St Martins Road,
Coventry,
UK

P. Holden and J.Q. Marchant

Anglian Water Ltd,
Thorpewood House,
Thorpewood,
Peterborough,
UK

1 ABSTRACT

Severn Trent and Anglian Water serve over 13 million customers in central and eastern England. Following recent droughts both companies see the River Trent as a potential new drinking water resource capable of safeguarding existing supplies in the East Midlands and Lincolnshire.

The River Trent, hitherto an 'untapped' resource for potable supplies, contains the treated sewage from 4 million people together with the influence of industrial discharges from the Midlands means that the river quality can be highly variable. In particular high levels of pesticides, flame retardents, boron, bromide and nitrate in the Trent render it difficult to treat using conventional water treatment processes. Alternative treatment options such as ultrafiltration and reverse osmosis have been identified and the results of preliminary trials used to develop a programme of pilot plant studies undertaken as a joint venture between Severn Trent and Anglian Water. Results to date compare the treatability of key risk compounds by conventional and membrane treatment processes and the effect on the formation of disinfection by-products at the elevated bromide levels often found in the Trent.

2 INTRODUCTION

The River Trent drains the large urban catchment of the Midlands. The industrial and domestic waste from Birmingham, Nottingham and Leicester, in addition to run-off from highways and agricultural activities, all contribute to the contamination of the River Trent. Historically the river has always been considered to be 'too polluted' for drinking water abstraction but, due to the improvement in quality of its tributaries and the tighter standards and compliance of sewage effluent, the river has recently undergone a marked quality improvement.

In the light of the above quality improvements and the projected need for security of water supplies in the East Midlands and Lincolnshire, the River Trent was identified as a potential source of potable water in early 1996. Severn Trent Water being interested in utilising Trent water at their existing works situated between Nottingham and Derby (Church Wilne WTW) and Anglian Water for a new works close to Lincoln. There was then a need to identify a suitable water quality monitoring and treatment strategy capable of producing final water to meet all current and potential future drinking water standards. This paper describes a series of pilot plant studies designed to compare conventional treatment of River Trent water with treatment using state of the art membrane technology.

3 DESIGN OF THE ADVANCED WATER TREATMENT PLANT

The Advanced Water Treatment Plant (AWTP) was designed following preliminary pre-treatment and Reverse Osmosis (RO) studies. Spiking trials on single RO elements were carried out to compare the rejection of Trent risk compounds (boron, bromide, nitrate and isoproturon) achieved by both High Rejection (HR) and Ultra Low Pressure (ULP) RO membranes. There was some deterioration of rejection achieved by the ULP RO (55 to 40%) for boron removal but all other spiking compounds were removed by >95%. ULP RO elements were therefore selected for the AWTP in view of the energy savings when compared with running HR elements. In pre-treatment pilot trials the performance of self cleaning, fine screen filters was compared with a self cleaning sand filter (Dynasand) prior to UF. Incoming turbidities between 1 and 120 NTU were experienced, with only the Dynasand filter producing filtrate of sufficient quality in all events.

The AWTP was commissioned in May 1998, and comprises a 350 m³/d 'Conventional' stream (based on the flowsheet of Church Wilne WTW) running parallel to a 350 m³/d 'Membrane' stream (*Figure* 1). Both streams can also make use of ozone and powdered activated carbon (PAC) and the modular nature of the individual processes allows them to be interchanged. Thus, the initial direct comparison will be followed by the optimisation of a process for the River Trent, possibly by combining processes from the two flowsheets.

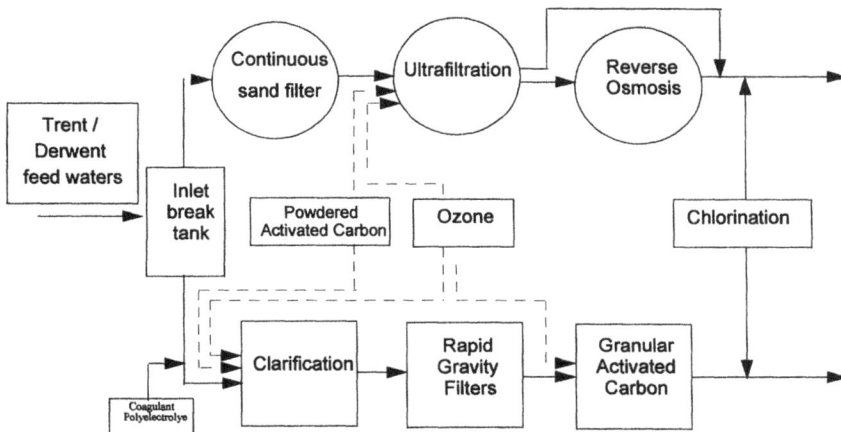

Figure 1 *Schematic of Advanced Water Treatment Plant*

The plant is monitored and to some extent controlled using an 'In-Touch' SCADA package, which also provides historical data trending and can be accessed remotely. Weekly analysis and spiking trials with problem determinands is testing the performance of the various processes, allowing an optimum treatment process for the River Trent water to be developed, based on final water quality and operating costs.

Feed waters to the Advanced Water Treatment Plant (AWTP) are available from four different sources:
- Direct abstraction from the River Trent
- Witches Oak Waters (River Trent from bankside storage)
- Church Wilne Reservoir (impounded Derwent)
- River Derwent (pumped from the Church Wilne WTW reservoir intake line)

The plant can treat a single feed water, or a blend of Trent and Derwent waters, mixed in the inlet break tank. This is important to both Water Companies involved as Severn Trent Water intend initially to treat a blended water from the two rivers and Anglian Water will take the Trent downstream of the Trent/Derwent confluence.

3.1 Conventional Treatment Processes

The conventional treatment stream is based on the Church Wilne WTW flowsheet and comprises clarification, followed by four rapid gravity sand filters (RGF) and four granular activated carbon (GAC) adsorbers.

The cone shaped upflow Clarifier is designed to process $15m^3/h$ of water from the inlet break tank, and uses ferric sulphate coagulant and a starch based polyelectrolyte as a coagulant aid. A height adjustable sludge cone collects the floc blanket and the clarified water overflows under gravity to the RGF header tank. Here it is distributed equally to the four 4.4m (height) x 1m (diameter) filters, each of which contains sand and anthracite on a gravel support. The RGF are backwashed every 48 hours or on head loss. The filtrate from all four filters enters a common line and flows to a break tank, which is used both to backwash the RGF and feed forward onto the GAC via a relift pump. The four 6m (height) x 1m (diameter) GAC columns are fed from a common header tank and designed to process a combined flow of $14m^3/h$, the depth of Chemviron F400 activated carbon in each adsorber resulting in a designed empty bed contact time of 30 minutes with all four units in use.

3.2 Membrane Treatment Stream

The membrane stream comprises a self cleaning sand filter (Dynasand), an ultrafiltration skid and a reverse osmosis skid

Raw water is pumped to the bottom of the DST-15 Dynasand and filters up through approximately 3m of 1.2-2mm sand, the filtrate flowing over a weir at the top. The sand is cleaned and recycled by means of an air lift pump running from top to bottom through the centre of the Dynasand. Air, injected at the lowest point of the unit, causes a water/sand/dirt mixture to rise to the top of the unit where it emerges in an open bottom vessel. Filtrate flows upwards into this vessel and escapes over a weir to waste, carrying the dirt with it, the sand meanwhile sinks back down on to the top of the sand bed. The Dynasand filtrate overflows to a break tank from which the Ultrafiltration skid draws its feed.

The Norit MT Ultrafiltration plant (containing X-flow Magnum hollow fibre membrane elements) is designed for a net daily production of 350m³, at a flux of approximately 80lm⁻²h⁻¹. The unit operates in dead-end mode and is taken off line automatically at pre-set time intervals to backwash or chemical clean (using sodium hypochlorite or nitric acid) to maintain a given trans-membrane pressure at >90% recovery. The filter material is polysulphone based with a nominal molecular cut off of 150-250kD and reduces the turbidity and suspended solids in the permeate to below the limits of detection. The permeate quality is therefore consistent, and any breach in the integrity of the unit can be identified by a detection of turbidity, in particle counts or an increase in the particle index. The UF permeate is collected in a 10m³ break tank, which is used to backwash the UF and feed the reverse osmosis skid.

The Reverse Osmosis skid contains fluid systems polyamide, spiral wound, TFC-ULP 4" membranes and operates in either 4:2:1 or 4:3 configuration. The unit is designed to treat 240m³/day at 80-90% recovery, and is dosed with antiscalant to prevent scaling on the membrane surface. Computer software is used to set the RO plant configuration, antiscalant dose and to assess its performance.

3.3 Ancillary Processes

The ozone plant has been used prior to UF and GAC treatments and is capable of introducing up to 6mg/l of ozone into a flow of 15m³/h with 3 minutes contact time.

The Powdered Activated Carbon plant has to date been used prior to clarification, dosing up to 40mg/l of Norit SA Super PAC. Use of PAC prior to UF will be included within the agreed programme of work.

4 RESULTS AND DISCUSSION

The programme of work at the AWTP includes routine sampling and analysis to optimise performance of each treatment process relative to the feed water quality, regular spiking trials of key risk compounds to ascertain cost effective removal and special projects such as disinfection by-product control.

Figure 2 shows the removal capability of GAC and RO treatments for high levels of key risk compounds actually found to date in River Trent feed water.

Boron being such a small element is unusual in that it is only removed to the same extent by RO as conventional treatment, approximately 50%. This could be important if the current UK limit of 1000µg/l was reduced to WHO guide level of 300µg/l. To date up to 60% removal has been seen as a result of pH adjustment to 8.5 prior to RO and this will be investigated further together with the effects of reducing pH.

Bromide was detected in River Trent feed water at concentrations between 150 and 540ug/l. It is removed by RO to its limit of detection but has a mean concentration of 280ug/l (maximum 450ug/l) post GAC. Bromide is not included as a regulated parameter within drinking water standards but it plays a significant role in the formation of trihalomethanes and bromate resulting from disinfection with chlorine and ozone treatment respectively.

Nitrate levels in the Trent can at times exceed the 50 mg/l limit which can not be reduced using the conventional processes currently available at Church Wilne WTW. To

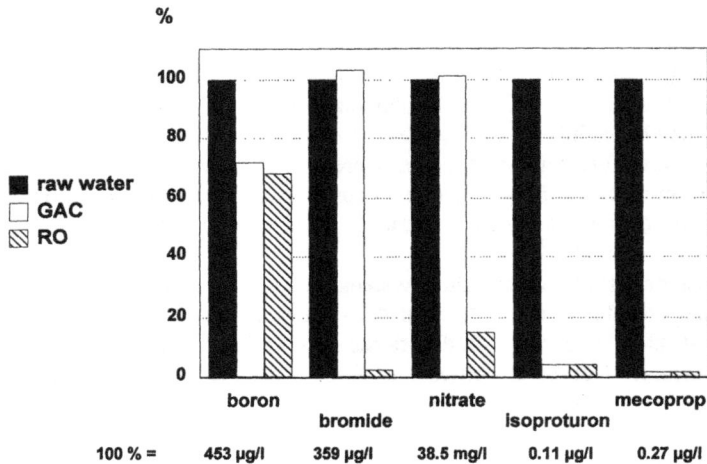

Figure 2 *River Trent AWTP Results - Occurrence of Key Risk Compounds*

reduce nitrate levels in the Trent to drinking water standards would require either for it to be blended 1:1 with River Derwent water (the current raw water source), or for it to be treated with another process such as Reverse Osmosis (RO). A combination of Ultrafiltration (UF) and RO (treating a part of the flow) can be considered as an alternative to blending and conventional treatment.

UF offers a 'total barrier' method of disinfection when considering removal of cryptosporidia, giardia, bacteria and viruses. Chlorine addition after UF is therefore only required to give a protective residual rather than full disinfection as is the case for the conventional stream. Micobiological data to date show the complete absence of bacterial or virus penetration of UF. The reported 3 log removal for cryptosporidia and giardia (oocysts and cysts) by UF will be checked within the work programme.

RO treatment achieves approximately 90% removal of Alkalinity and Total Hardness (raw feed waters being typically 300 and 150 mg/l as $CaCO_3$ respectively). This yields a slightly acidic, weakly buffered product requiring stabilisation by remineralisation or blending with UF or another treated supply to be suitable for public supply. For cost effective treatment to drinking water standards a minimum percentage of the flow should be treated by RO and the need for remineralistion be avoided if possible.

Severn Trent have established a very advanced on-line monitoring unit on the bankside of the River Trent, to establish the necessary river intake protection to the proposed water supply scheme. The on-line monitors includes a SAMOS-LC system using a conventional HPLC technique on-line for the analysis of organic pollutants. This monitor originally developed for the analysis of triazine and phenylurea herbicides in the River Rhine system is unique in this River Trent installation in that it includes a modification to allow the detection of several acid herbicides and phenolic compounds known to be frequently present in the River Trent. Detection of three such acid herbicides in July 1998 in the River Trent at approximately 0.3 µg/l resulted in sampling of RO permeate and GAC product waters. The complete elimination of these acid herbicides to below analytical limits of detection by these treatment processes was observed. *Figure* 2 shows the removal capability of GAC and RO treatments for isoproturon and mecoprop when treating 0.1 and 0.27 µg/l concentration respectively.

A wide range of herbicides (triazines, urons and acid herbicides) have been spiked at 2 µg/l with GAC treatment of 15 minutes empty bed contact time (EBCT) and RO achieving < 0.1µg/l in all cases. This matches the capability of the on-line pesticide monitoring facility which alarms when selected herbicides from each group exceed 1.0 µg/l. Ozone doses of 2 mg/l applied reduced the pesticide loading on to GAC or RO by between 25 to 50 % across the range of herbicides added although ozone residuals after 3 minutes contact had to be avoided to prevent the formation of bromate. PAC (at 20 mg/l) has been seen to reduce 4 µg/l by 80% for most of the above range of pesticides.

The flame retardents TNBP, TCEP and TCPP have been regularly spiked at 2.0, 5.0 and 10 µg/l respectively in line with the maximum concentrations typically found in Trent and Derwent derived feed waters. TNBP and TCEP are removed to <0.1 µg/l by GAC (15 minutes EBCT) and RO. TCPP is reduced to <0.5 and <0.1 µg/l respectively by GAC and RO treatment. Ozone treatment did not show any effect on the three common flame retardents selected. Severn Trent Water currently has an internal standard of 1 µg/l for individual flame retardents, should tighter drinking water standards be set than 0.5 µg/l in the future then total flow through RO or a combination of GAC/RO or PAC/RO would be required.

Two chlorine contact rigs complete the treatment for each pilot plant treatment stream. Data collected from these rigs to date confirm the consequences of high Br and TOC during summer on THM levels found from bench scale trials. Results show that at 300 µg/l bromide there is a shift to brominated haloforms, giving the potential to exceed the 100 µg/l THM limit by both GAC or UF treated water at about 350 µg/l bromide concentration as seen in Table 1. The conditions used for each test was 1.0 mg/l chlorine residual was maintained for 1 hour then samples were stored for 48 hours at 22°C. Chlorinated RO water shows almost no THM production hence a 1:1 blend of UF and RO gives 50% of the UF value. RO treatment may become important if tighter THM standards are set in the future.

For conventional treatment of a Trent :Derwent blend raw water bromide must be restricted to a maximum of 300 µg/l. On-line monitoring facilities are in place to ensure Trent bankside storage is keep within the required limits.

Table 1 *AWTP - Effect of bromide concentration on Trent water THM formation*

Treatment	Total THM at 300µg/l Br	Total THM at 400µg/l Br
UF	95	115
GAC	85	105
1:1 UF+RO	48	55

PAC has been shown to reduce THM formation potential on the conventional plant by approximately 33% at 200µg/l Br despite a significant shift towards forming brominated derivatives. The benefits of ozone for THM control have been found to be minimal for either stream.

Operational cost comparisons calculated from AWTP data indicate that 100% treatment by UF followed by 50% RO treatment would be 1.5 to 2.5 times as expensive as conventional treatment (utilising 15 minutes EBCT), however the membrane stream has yet to be optimised.

In summary conventional treatment in conjunction with on-line low level pesticide protection can treat a blend of Trent : Derwent water to potable standards (with the possible exception of THM exceedances due to high bromide levels), whilst the membrane stream is capable of treating River Trent water directly but may require a high percentage pass through RO.

5 CONCLUSIONS

High concentrations of pesticides, flame retardents, boron, bromide and nitrate, and the nature of their occurrence in the Trent render it difficult to treat using conventional water treatment processes to current and potential future UK drinking water standards.

On-line monitoring of bromide and low level pesticides can be used in conjunction with bankside storage to reduce treatment costs by limiting the challenge of key risks compounds to predetermined maximum concentrations.

A membrane stream has a significantly smaller footprint (approximately 25%) than that needed for conventional processes minimising civil engineering requirements.

The increase in operational cost of the membrane processes when using 50% pass through RO is estimated at 1.5 to 2.5 times that of conventional treatment however this factor is likely to reduce as UF and RO processes are optimised and developed.

A hybrid system using conventional and membrane plant may be the cost effective answer for the Trent.

Hollow Fibre Ultrafiltration Membrane Skid (Conventional Stream in Background)

Ultra-Low Pressure Reverse Osmosis Membrane Treatment Unit

THE USE OF ELECTRODIALYSIS AT AMSTERDAM WATER SUPPLY

J.P. van der Hoek[1], J.A.M.H. Hofman[1], P.A.C. Bonné[1] and D.O. Rijnbende[2]

[1]Amsterdam Water Supply,
Vogelenzangseweg 21,
2114 BA Vogelenzang,
The Netherlands

[2]Ionics Nederland,
Hollandsch Diep 69,
2904 EP Capelle aan den IJssel,
The Netherlands

1 INTRODUCTION

Amsterdam Water Supply has a total drinking water production capacity of 101 million m^3/year, divided over two production plants. Production plant Leiduin has a capacity of 70 million m^3/year and research has been carried out into extension of the capacity of this plant to 83 million m^3/year. The extension concerns the use of an Integrated Membrane System (IMS), which will be incorporated in the existing scheme. Figure 1 shows the existing scheme and three alternative IMSs for extension.

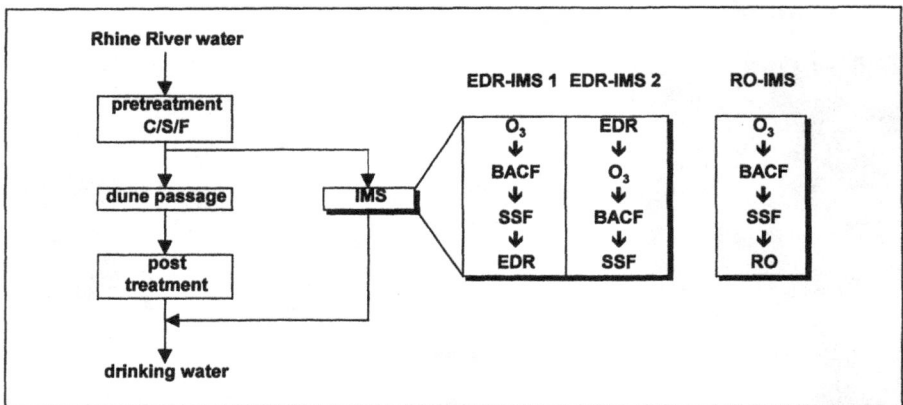

Figure 1 *Alternatives for extension of the drinking water production plant "Leiduin" with an Integrated Membrane System*

In the existing scheme Rhine River water is pre-treated by coagulation-sedimentation-filtration and after dune passage post-treated by rapid sand filtration, ozonation, pellet softening, biological activated carbon filtration and slow sand filtration. In the extension schemes, the pre-treated Rhine River water is directly treated without soil passage by an IMS. The IMS comprises either reverse osmosis (RO) or electrodialysis (EDR). Because in the existing plant the process unit's ozonation, biological activated carbon filtration and slow sand filtration already have a capacity of 83 million m³/year, these units are used in the IMS as pre- or post-treatment for the membrane system. Treatment goals of the IMS are desalination, softening, removal of organics and disinfection. In addition the system has to be cost-effective, reliable and has to result in a low environmental burden. The IMS comprising RO has shown to fulfil all these requirements[1]. EDR has become an alternative for RO in the IMS because disinfection and removal of organics can be fully guaranteed by the ozonation, biological activated carbon filtration and slow sand filtration. Hence, the membrane system has to result only in desalination and softening[2].

2 APPLICATION OF EDR IN AN IMS

Using EDR in the IMS, it can be placed either as final process unit or as first process unit. Figure 1 shows these possibilities. In addition, the performance of EDR can be affected by staging. In the research, a 2 stage and 3 stage EDR units were compared used as first process unit as well as final process unit in the IMS. The following items were incorporated in the research to find the optimum IMS configuration and EDR system:

1. The desalination characteristics of a 2 and 3 stage EDR system. The treatment goal was to reach desalination and softening of at least 80%. As the temperature of the pre-treated Rhine River water varies between 0 °C and 25 °C, also the effect of temperature on the desalination characteristics was studied;
2. The disinfection strategy in relation with bromate formation. In the IMS using EDR, disinfection relies for the major part on ozonation, as EDR has no disinfection credits. To comply with the disinfection requirements, the ozonation has to result in a CT value of at least 15 mg.min/l, necessitating an ozone dose of 2-3 mg/l. Due to the low DOC content of the pre-treated Rhine River water (2 mg/l) and the high bromide content (150-300 µg/l), bromate will be produced, and the Dutch standard of 5 µg/l may be exceeded. Bromate has to be removed by EDR in process scheme EDR-IMS 1, or the precursor of bromate, bromide, has to be removed prior to ozonation in process scheme EDR-IMS 2 to reduce the bromate formation potential;
3. Fouling of the EDR system. In scheme EDR-IMS 1 fouling is no problem due to the excellent feedwater quality after the excessive pretreatment[1]. In scheme EDR-IMS 2 fouling will be more severe and a cleaning strategy has to be developed;
4. Acid consumption and energy consumption. The use of more stages in an EDR system will increase the use of chemicals and energy. The 2 stage and 3 stage EDR systems were compared on this item.

3 EDR PILOT-PLANT FACILITIES

Ionics built the 2 stage EDR unit as a new EDR 2020 system. Figure 2 shows the system. It consisted of two hydraulic and two electrical stages, each containing 250 cell pairs. The anion membranes were of type AR204-SZRA and the cation membranes were of type CR67-HMR, both manufactured by Ionics. The flowspacers were the new Mark IV (thickness 0.76-mm, size 46x102 cm, with an effective area 3450 cm²). The product flow was set at 14 m³/h by a variable speed drive controlling the feed pump. The concentrate recirculation flow was controlled to a differential pressure of 0.05 bar diluate overpressure to prevent cross-leaking of concentrate to diluate. Phased reversal and 4-way valves were applied to minimise the amount of off-spec product during reversal (reversal time 20-30 min).

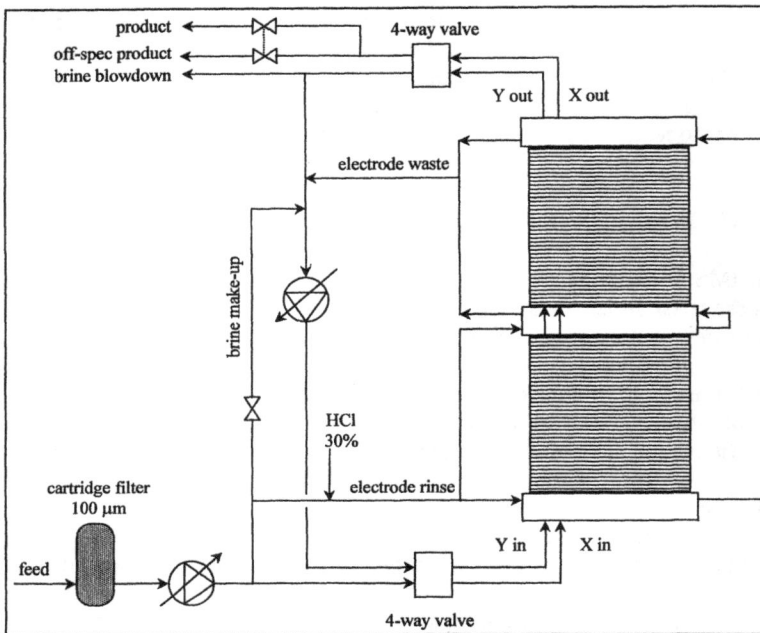

Figure 2 *Process scheme of the 2 stage EDR pilot-plant*

The 3 stage EDR system was built by modifying the 2 stage system, as shown in Figure 3. The two electrical stages were maintained, but the first electrical stage was divided into two hydraulic stages. So, the first electrical stage consisted of two hydraulic stages of 175 and 150 cell pairs respectively, and the second electrical stage consisted of one hydraulic stage of 175 cell pairs. The lower number of cell pairs in the second hydraulic stage of the first electrical stage was necessary to avoid polarisation in the second hydraulic stage. As these hydraulic stages form one electrical stage, they have the same current density. The salt content in the second hydraulic stage however is lower and thus polarisation may occur. The lower number of cell pairs in the second hydraulic stage of the first electrical stage results in a higher flow velocity, increasing the limiting current and thus avoiding polarisation.

The process conditions of the 2 stage and 3 stage EDR systems are summarised in table 1. In both cases hydrochloric acid (30%) was dosed to the concentrate to obtain a LSI of 1.2-1.5 to avoid scaling in the concentrate cells.

4 RESULTS AND DISCUSSION

4.1 Desalination characteristics

The desalination expressed as decrease of electric conductivity (EC) is shown in Figure 4 for the 2 stage and 3 stage EDR system.

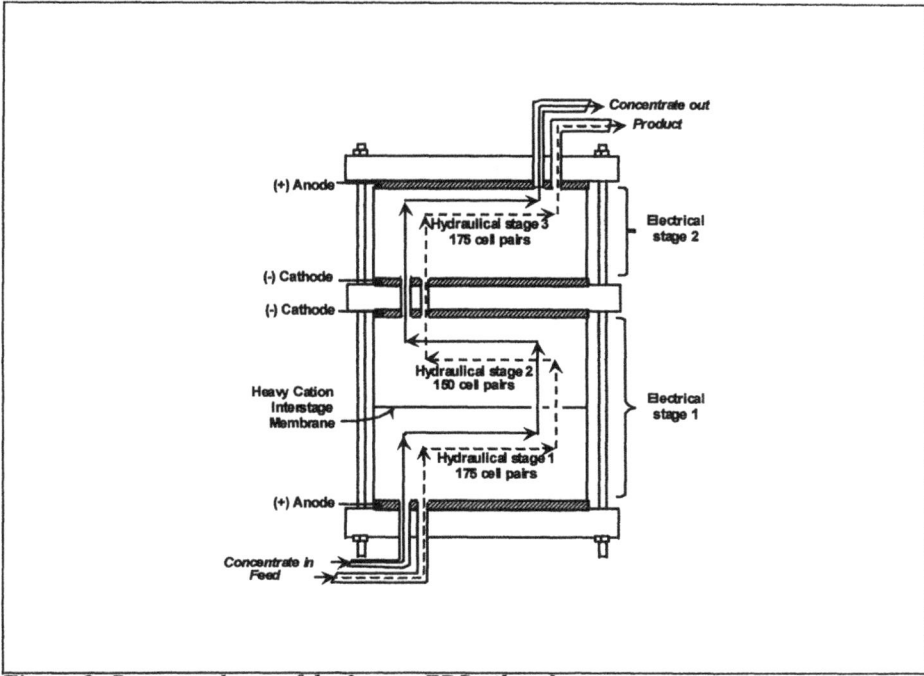

Figure 3 *Process scheme of the 3 stage EDR pilot-plant*

Table 1 *Process conditions of the EDR pilot-plants*

	2 stage EDR		3 stage EDR		
	Stage 1	Stage 2	Stage 1	Stage 2	Stage 3
No. of cell pairs	250	250	175	150	175
Voltage (V)	190	180	175	175	115
Polarisation (%)	60-80	60-80	40	70	85
Current efficiency (%)	90-95		90-95		
Product flow (m³/h)	14		8		
Recovery (%)	92-93		87-89		
LSI concentrate	1.2-1.5		1.2-1.5		

As can be expected, a 3 stage EDR system results in a higher desalination as compared with a 2 stage EDR system. For both systems the desalination reduces as the temperature decreases. The temperature dependency is approximately 0.8%/°C. The specific removal of chloride, sulphate and hardness for both systems is summarised in table 2. The results show that the 3 stage EDR system results in the required desalination of 80% over the whole temperature range. The 2 stage EDR system could only fulfil this requirement at a temperature of 20 °C or higher.

4.2 Disinfection strategy and bromate control

In the IMS comprising EDR disinfection is achieved for the major part by ozonation.

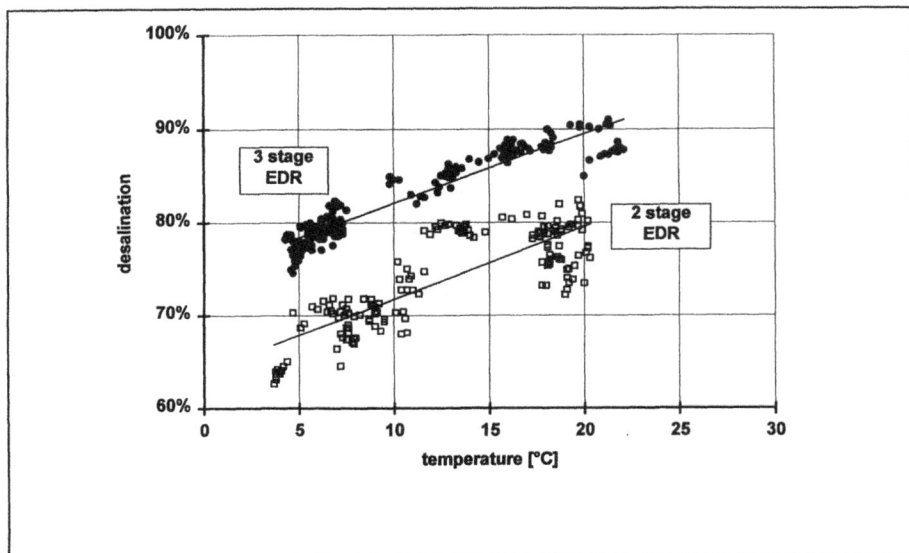

Figure 4 *Desalination characteristics of the 2 and 3 stage EDR system*

Table 2 *Removal of chloride, sulphate and hardness by the 2 and 3 stage EDR system*

Removal %	2 stage EDR		3 stage EDR	
	4-6 °C	22-24 °C	4-6 °C	22-24 °C
Cl⁻	72	83	83	91
SO₄²⁻	73	89	84	93
hardness	78	89	88	97

In order to have a robust disinfection barrier the ozonation has to result in a CT value of approximately 15-20 mg.min/l. From Figure 5 it can be seen that this CT value can be obtained with an ozone dose of 2-2.5 mg/l. However, ozonation of the pre-treated Rhine River water, containing 2 mg/l DOC and 150-300 µg/l Br⁻, will result in the formation of bromate in a concentration of 40-80 µg/l, as can be seen in Figure 6. The Dutch standard for bromate is 5 µg/l. Both EDR-IMS process schemes were judged on their efficacy to comply with this bromate standard.

4.2.1 EDR as final process unit in the IMS. EDR, following the ozonation step, can be used to remove the bromate formed during ozonation. In order to reduce the bromate concentration from 40-80 µg/l to 5 µg/l, the bromate removal has to exceed 90%. From experiments with the 2 stage EDR unit it was concluded that the bromate removal was restricted to 64% at a temperature of 2.4 °C. Both the first stage and second stage showed a bromate removal of 40%. In the IMS with EDR as final process unit the 3 stage EDR system was not tested. Assuming that the third stage would also result in a 40% bromate removal, a 3 stage EDR system should theoretically result in a bromate retention of 78%, still below the required 90%. So, it can be concluded that the IMS in which ozonation precedes EDR is not viable with respect to disinfection strategy and bromate control.

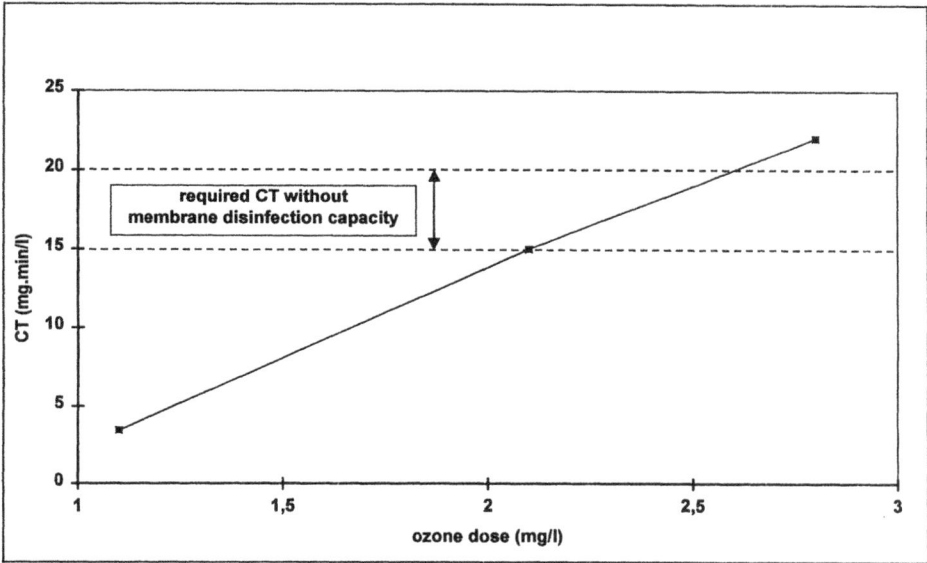

Figure 5 *Required ozone dose to meet the required CT value in the EDR-IMS*

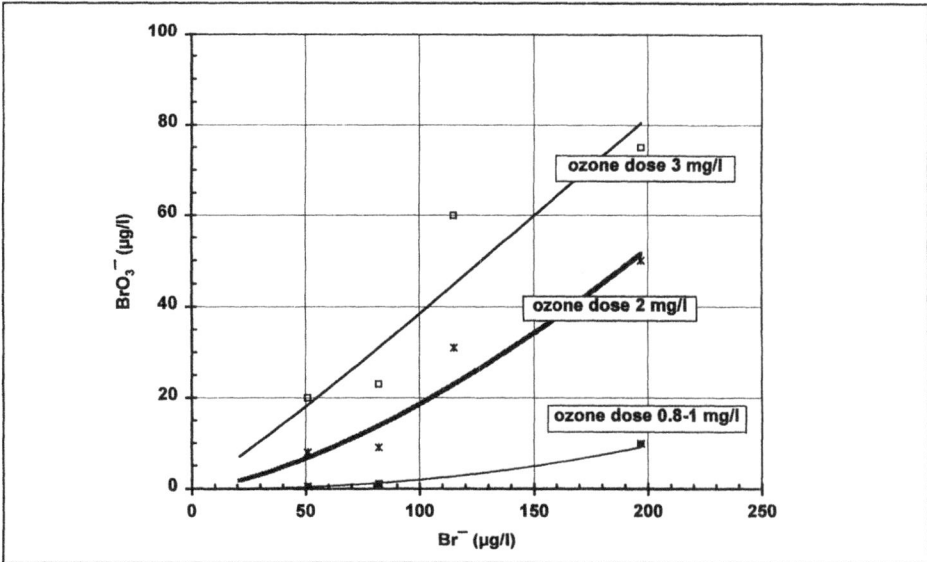

Figure 6 *Bromate formation as a function of ozone dose and bromide concentration (DOC content 2 mg/l)*

4.2.2 EDR as first process unit in the IMS. Applying EDR as first process unit in the IMS, EDR can remove bromide from the pre-treated Rhine River water prior to ozonation, reducing the bromate formation potential. In addition, the pH can be lowered during ozonation, a well-known method to reduce bromate formation. A pH decrease of 1-2 units will reduce the bromate formation with 50% (4,5). Figure 6 shows that the bromide concentration has to be reduced to approximately 30 µg/l to restrict the bromate formation to 5 µg/l at an ozone dose of 2-3 mg/l. With an initial bromide concentration of 150-300 µg/l in the pre-treated Rhine River water, bromide removal should be at least 80-90%. Figure 7 shows the actual bromide removal for the 2 stage and 3 stage EDR system.

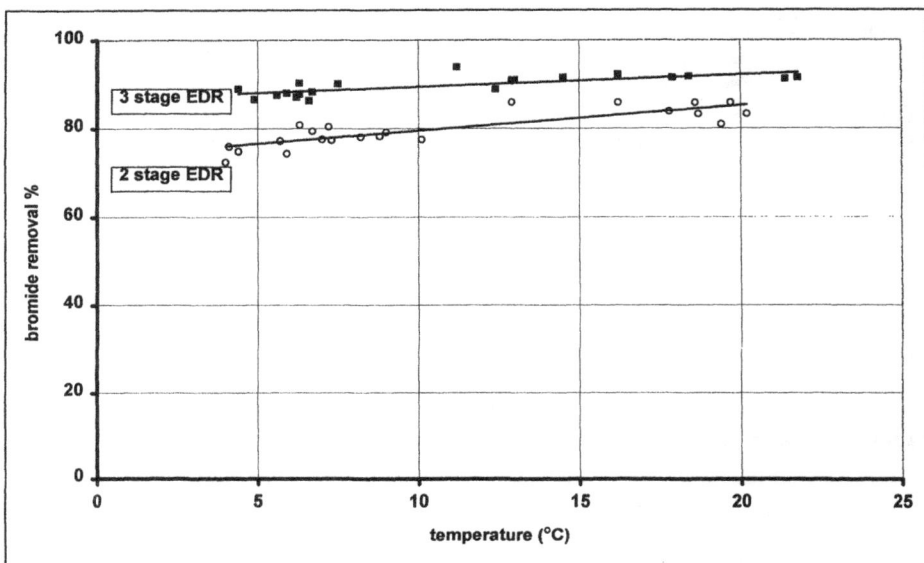

Figure 7 *Bromide removal with a 2 stage and 3 stage EDR system*

The 2 stage EDR system showed a bromide removal ranging from 75% to 85%. This means that especially at low temperatures bromide removal will not be sufficient to restrict the bromate concentration to 5 µg/l, and in addition acid has to be dosed during the ozonation to comply with the bromate standard of 5 µg/l. At a low water temperature (4 °C) the relationship between ozone dose, pH and bromate formation was established for the water treated with the 2 stage EDR system. Figure 8 shows the results. By lowering the pH to 6.5 the bromate formation can be restricted to 5 µg/l, even at an ozone dose of 2.3 mg/l.

The 3 stage EDR system showed a bromide removal ranging from 90% to 95% (Figure 7). This implies that bromide removal is sufficient to meet the bromate standard of 5 µg/l during ozonation. Under all circumstances, even without an acid dose, the bromate concentration after ozonation was in compliance with the standard (results not shown).

From these results it is clear that in EDR-IMS 2, applying EDR as first process unit, both with a 2 stage and 3 stage EDR system the bromate requirement can be fulfilled, albeit that with a 2 stage EDR system an additional acid dose is required. The acid dose (30% HCl) to control the bromate formation is 30 g/m^3 $_{product}$.

4.3 Fouling of the EDR system

Scaling and fouling can seriously affect the EDR performance. Scaling can be controlled by dosing HCl in the concentrate up to a LSI of 1.2-1.5, by restricting the recovery to 93% and by using a reversal time of 20-30 min.

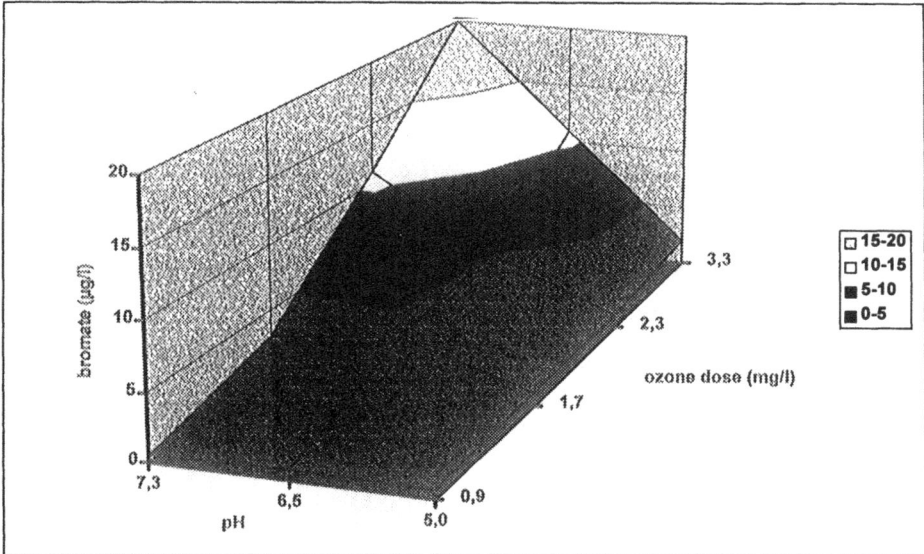

Figure 8 *Relation between ozone dose, pH and bromate formation for ozonation of the product water of the 2 stage EDR system*

To avoid fouling, Ionics gives guide values for the EDR feedwater and advises to use cartridge filters preceding the EDR unit. The Ionics guidelines and the actual concentrations in the EDR feedwater in scheme EDR-IMS 2 are compared in table 3.

Table 3 *Guidelines of the EDR feedwater quality and the actual feedwater quality (pre-treated Rhine River water) in EDR-IMS 2*

Parameter	Ionics guidelines	Actual concentrations
pH	5-9	7.6
Turbidity (NTU)	<1	0.10
SDI	<10	2.5
Fe (mg/l)	<0.3	<0.05
Mn (mg/l)	<0.1	<0.002
DOC (mg/l)	<10	2.2
Cl_2 (mg/l)	<0.5	0

In principle, no problems have to be feared in combination with the use of 100 μm cartridge filters. However, a periodic Clean In Place (CIP) acid cleaning of the stacks appeared to be necessary in EDR-IMS 2 to maintain the desalination capacity. A stable operation could be maintained by applying this CIP monthly, as shown in Figure 9.

4.4 Energy and chemicals consumption

Besides the desalination characteristics, an important difference between a 2 stage and 3 stage EDR system is the energy and chemicals consumption.

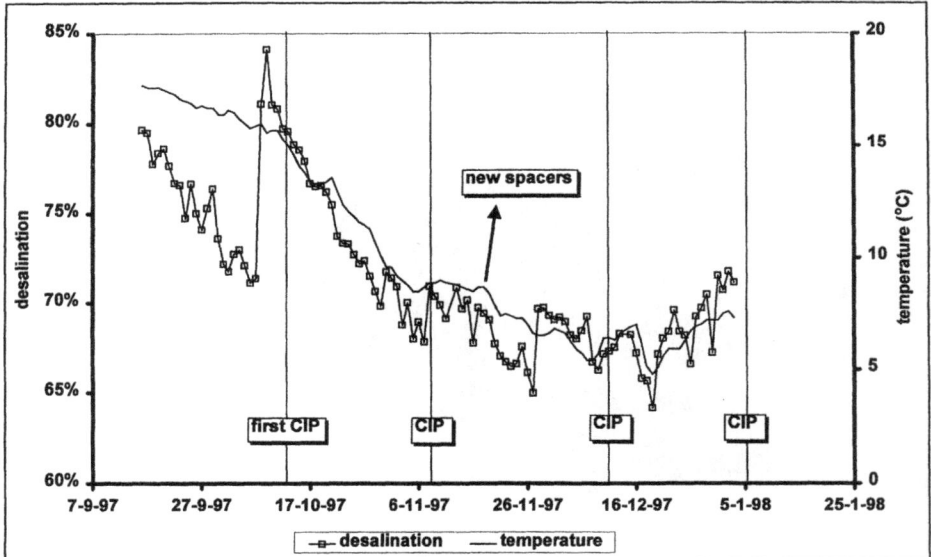

Figure 9 *Effect of CIPs (Clean In Place) on the desalination performance of EDR in EDR-IMS 2(2 stage EDR system)*

A 3 stage EDR system requires more energy and a higher acid dose for scaling control. On the other hand, in scheme EDR-IMS 2 a 3 stage EDR system does not need an acid dose for bromate control. The overall energy consumption (desalination energy and pumping energy) and the acid consumption of the 2 stage and 3 stage EDR system in EDR-IMS 2 are summarised in table 4.

Table 4 *Energy and chemicals consumption of a 2 stage and 3 stage EDR system in EDR-IMS 2*

	2 stage EDR	3 stage EDR
Energy consumption (kWh/$m^3_{product}$)	0.4	0.6
Scaling control (g/$m^3_{product}$ 30% HCl)	29	35
Bromate control (g/$m^3_{product}$ 30% HCl)	60	0

5 CONCLUSIONS

Extension of the capacity of one of the production plants of AWS is possible with an IMS, comprising ozonation, biological activated carbon filtration, slow sand filtration and EDR. To reach the required desalination degree ($\geq$80%), only a 3 stage EDR system can be used in the IMS. To reduce the bromate concentration formed during ozonation to the standard of 5 μg/l; the EDR system should be used as first process unit in the IMS. The removal of bromide by EDR results in a reduced bromate formation potential. A 3 stage EDR system has the advantage above a 2 stage EDR system that under all circumstances the bromate standard can be met. Using a 2 stage EDR system, an additional acid dose is required to control the bromate formation. Fouling of the EDR system can be easily controlled by a monthly CIP.

The energy consumption of a 3 stage EDR system is higher than the energy consumption of a 2 stage EDR system, but comparable with RO (6). Although the acid consumption for scaling control in a 3 stage EDR system is higher than in a 2 stage EDR system, the overall acid consumption of EDR-IMS 2, using a 3 stage EDR system, is about 50% as compared with EDR-IMS 2 using a 2 stage EDR system, as no acid is required for bromate control.

The environmental burden of EDR is low as compared with RO: the concentrate volume is 50%-66% as compared with RO, while the chemicals consumption is six times lower as compared with RO (7).

The 3 stage EDR system is characterised by 40% higher investment costs as compared with the 2 stage EDR system, but in the IMS it can compete with RO.

Using an IMS consisting of subsequently EDR, ozonation, biological activated carbon filtration and slow sand filtration, it can be concluded that a 3 stage EDR system applied as first process step results in the optimum process configuration. In that case the EDR-IMS can fulfil all requirements with respect to desalination, softening, removal of organics and disinfection.

References

1. J.P. van der Hoek, J.A.M.H. Hofman, P.A.C. Bonné, M.M. Nederlof and H.S. Vrouwenvelder, *Desalination*, 1999 (in press).
2. J.P. van der Hoek, D.O. Rijnbende, C.J.A. Lokin, P.A.C. Bonné, M.T. Loonen and J.A.M.H. Hofman, *Desalination*, 1998, **117**, 159.
3. J.P. van der Hoek, P.A.C. Bonné, E.A.M. van Soest and A. Graveland, *Proceedings of the 21ˢᵗ IWSA Congress & Exhibition*, 1998, **21**, SS1-11.
4. G. Amy, M. Siddiqui, K. Ozekin and P. Westerhoff, *Wat. Supply*, 1995, **13(1)**, 157.
5. R. Sorg, P. Westerhoff, R. Minear and G. Amy, *Journal AWWA*, 1997, **89(6)**, 69.
6. J.P. van der Hoek, in *Alternatives for extension of the production capacity of Amsterdam Water Supply*, report Amsterdam Water Supply, 1999, Chapter 5.5, p. 81 (in Dutch).
7. J.P. van der Hoek, J.A.M.H. Hofman, P.A.C. Bonné and D.O. Rijnbende, *H₂O*, 1999, **32** (in press) (in Dutch).

APPLYING ELECTRODIALYSIS (EDR) TECHNOLOGY TO UNDERGROUND WATER TREATMENT

By Dr. Elis Sgarbi - Analysis Laboratory Department

Azienda Multiservizi Intercomunale - Consorzio di Imola
Via C. Casalegno n. 1,
40026 Imola (BO)
Italy - tel. 0542/621.232

1   FOREWORD

Drinking water from underground sources is still largely prevalent in Italy compared to surface sources (the former covers 87.5%, the latter 12.5%, according to an ISTAT study in 1987). The quality of the underground water coming from wells of various depths is of course affected by the geochemical make-up of the ground, but also - and very much so - by human activities on the imbriferous basin feeding the water table that the wells are drawn from. One quality problem with underground water has by now been proven to be strongly affected by human activity (agriculture, industry and town life) - pollution by $NO3^-$ nitric ions. The MAC for this was set in the EEC Directive 778/80 (adopted in Italy as D.P.R. 236/88) at 50 mg/l. The serious nature of the nitrate problem in Italy is due not only to the MAC being overstepped in limited individual areas (where the concentration may be as high as 150 mg/l), but especially to the extent of the territory involved, where the stratum used for drinking water is progressively rising (Marche, Emilia Romagna, Lombardy and Veneto). It is no coincidence that this is one of the most densely populated areas of Italy, and uses about one third of all drinking water in the country.   The territory where Consorzio A.M.I. draws its underground water is no exception as far as the nitrate issue is concerned, although it has limited peaks that rarely exceed concentrations of 70 mg/l. However, the excellent overall quality of the water drawn has led A.M.I. ever since 1990 to look for nitrate treatment technology solutions which would alter the natural features of the water as little as possible, while providing good performance in terms of nitrate ion abatement, high percentages of recovery and low power consumption.

2   THE DECISION TO USE POLARITY REVERSAL ELECTRODIALYSIS TECHNOLOGY (EDR)

A.M.I. first used electrodialysis technology (henceforth abbreviated as EDR) in 1990. At the time, the decision to adopt this kind of technology (which has rarely been applied to water supply systems anywhere in the world, even less so in Europe[1]) was inspired by its intrinsic features and by its many applications in the food industry. Generally speaking (Figure 1), EDR belongs to a range of technologies which use membranes to separate

ELECTRODIALYSIS

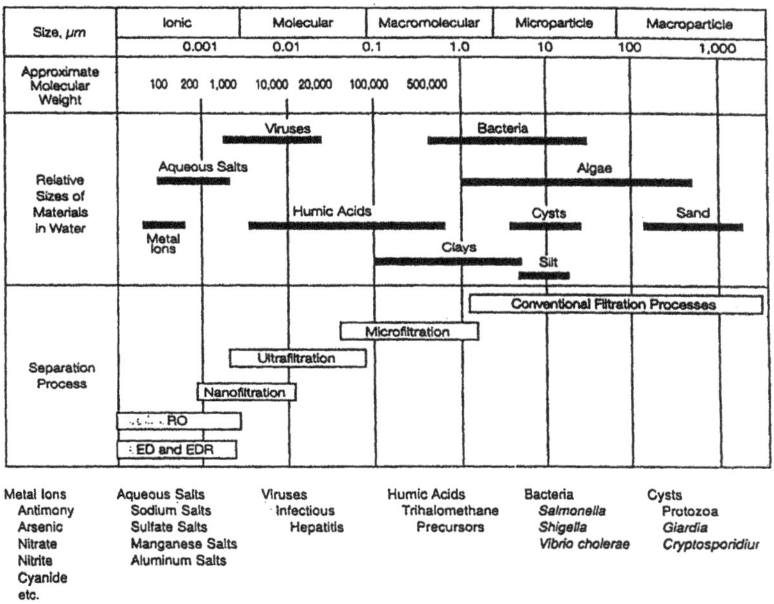

Size, μm	Ionic		Molecular		Macromolecular		Microparticle		Macroparticle	
	0.001		0.01		0.1	1.0	10	100	1,000	

Figure 1 reproduced as table/diagram:

Approximate Molecular Weight	100 200 1,000	10,000 20,000 100,000 500,000				
Relative Sizes of Materials in Water	Aqueous Salts / Metal Ions	Viruses / Humic Acids / Clays	Bacteria / Algae / Cysts / Silt	Sand		
Separation Process	RO / ED and EDR	Nanofiltration / Ultrafiltration	Microfiltration	Conventional Filtration Processes		

Metal Ions	Aqueous Salts	Viruses	Humic Acids	Bacteria	Cysts
Antimony	Sodium Salts	Infectious	Trihalomethane	*Salmonella*	Protozoa
Arsenic	Sulfate Salts	Hepatitis	Precursors	*Shigella*	*Giardia*
Nitrate	Manganese Salts			*Vibrio cholerae*	*Cryptosporidiur*
Nitrite	Aluminum Salts				
Cyanide					
etc.					

Figure 1 *Membrane Process Overview*

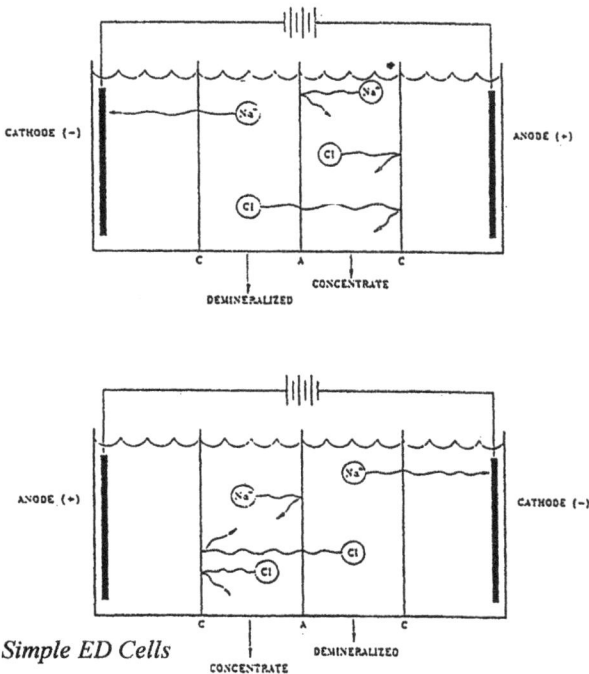

Figure 2 *Simple ED Cells*

ionic and non-ionic substances from a main flow, concentrating them in a so-called concentrated flow (microfiltration, ultrafiltration, nanofiltration, reverse osmosis and electrodialysis). Especially, EDR exploits the principle of the passage of current through ionic solutions in order to separate salts or certain ions from a diluted to a concentrated flow, through the use of so called permselective membranes.

It is a well-known fact that a current (carried by all the ions present) passes through the solution in an electrolytic cell containing a totally dissociated salt (like NaCl) under the effect of a continuous electric field: the cations migrate towards the cathode and the anions towards the anode. If we introduce at least three membranes of different selectivity into the cell, so that the cell is subdivided into four compartments, then the ions will be freer to move through the solution, following the direction of the electric field applied, because of the barriers set by the membranes (Figure 2). The layout of the electric field compared to the membranes leads to the electric current making the positive ions migrate towards the cathode and the negative ones towards the anode, inducing electrolyte impoverishment in one compartment and enrichment in another. Electrodialysis therefore has developed the so-called membrane battery system, i.e. a succession of anionic and cationic membranes separated from each other by suitable spacers (made of inert plastic material), thus creating a repetitive system of pairs of cells, each made up of:
a)   a cationic membrane;
b)   a demineralised water flow separator;
c)   an anionic membrane; d) a concentrated water flow separator.

Suitable re-engineering has led to the industrial production of membrane packets consisting of a variable number of cell pairs (300 to 500 pairs). An especially important point is that the flow of the fluid inside the packet or battery of membranes does not pass vertically through the battery, but horizontally; therefore the two fluid currents (the demineralised flow and the concentrated flow) move inside the electrolytic cells at a speed which - pressure being the same - is affected only by the capacity or flow speed (Figure 3). One of the most interesting features of this technology therefore is that one can work at low pressure levels (1 to 3 bar), hence with intrinsic low energy consumption. It is also possible, within certain limits, to increase the flow to the membrane packet by increasing the number of cell pairs, without being forced to increase the working pressure. For further details on specific equations governing the electrodialysis parameters, see the specialised bibliography[2] . Generally speaking, membranes are the field where the greatest developments are expected in the short and medium term. In the specific field of production of membranes for electrodialysis, larger sized membrane sheets have been developed which offer the same mechanical stability, but larger contact surfaces.

This makes it possible to increase the supply flow to the membrane packet while keeping the same number of cells. Practically speaking, this increases the size of the battery in two directions only (this is a very important condition for limiting space, especially in the case of systems hosted in containers).

3 FEATURES OF A.M.I.'s EDR SYSTEM

The treatment system employed by A.M.I. to reduce the concentration of nitrate in underground water is of an Electrodialysis Reversal type, styled by its initials EDR. The system - the main features of which are shown on Table 1 - has been connected to one of the artesian wells which draw water from the underground water table (depth about 100 m) in the area of the conoid of the river Santerno, in the Township of Imola (fig. 4). This well, like all the others of the same kind from which water is drawn to feed the water system of

EDR FLOW DIAGRAM OF A.M.I. PLANT

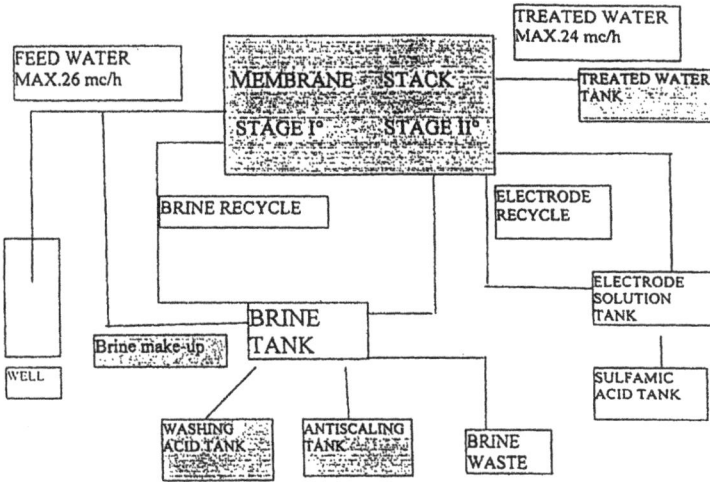

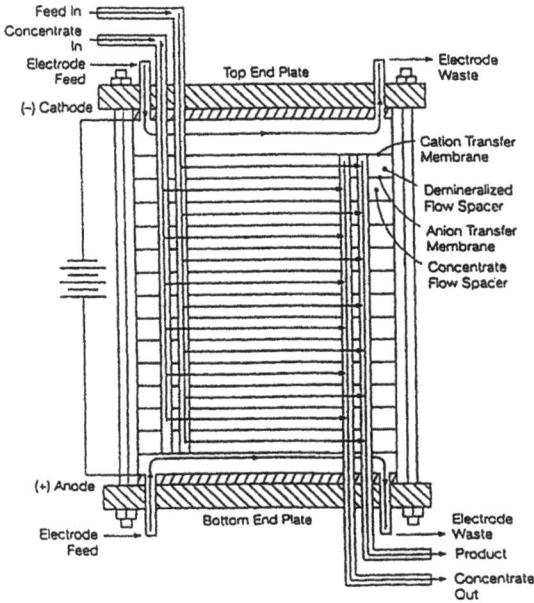

Figure 3 *Membrane-stack flow paths*

the Township of Imola, has general quality features - and especially micro-biological features - which make traditional disinfecting treatment unnecessary. The system has a limited throughput, but has all the features of a true productive installation. Also, since it was designed as a pilot plant, it has been provided with a data acquisition system on a PC connected to the automatic management PLC of the system; this PLC also acts as a service system for automatic collection of water samples from the three main flows (incoming water, treated water and concentrated brine. The average daily samples or the instant samples picked up from the three flows are then submitted to chemical and chemical/physical analyses at the company Analysis Laboratory. As already mentioned, the system is provided with a management PLC which automatically governs each set-up stage of the machine cycles: polarity reversal of the electric field, valve opening and closing, membrane packet washing phases, additive proportioning, plant alarms and down time, auxiliary utilities, etc.). To follow the stages of the job, the power cabinet is equipped with an overview panel showing the main machine conditions (solenoid valve condition, pump condition, measurements of pressure, temperature, conductivity, etc.).

The same power cabinet also hosts a manual control panel used under non-productive conditions to perform special control manoeuvres, especially in case of unscheduled washing. The PC installed inside the container and connected to the PLC is used not only to acquire data, but also for various programming modifications on the PLC (washing time, polarity reversal time, sampler programming, etc.).

Table 1 *Characteristic Data Of The A.M.I. Imola Edr Plant*

Container size	L 6050 X W 2430 X H 2800 mm
Potential throughput	24 m3/h of treated water
Installed electric power	20 kW
Membrane stack	Mod. EUR 20 B 850 (EURODIA-FRANCE)
Number of treatment stages-lines	Stages No. 2 - lines No. 1
Number of electrodes	4 for each stage (graphite type)
Total no. of cells	850
Anionic membranes	Neosepta mod. ACS (Tokuyama Soda)
Cationic membranes	Neosepta mod. CMX (Tokuyama Soda)
Actual membrane area	1683 cm2
Centrifugal pumps	No. 3
Brine tank	No. 1 for 300 lt.
Electrolyte tank	No. 1 for 100 lt.
Washing acid tank	No. 1 for 70 lt.
Additive tank	No. 2 for 70 lt. Each
Proportioning pumps	No. 3 (centrifugal type for acid, membrane type for additives)
Input filtration unit	No. 1 sleeve type
Electrolyte filtration unit	No. 1 sleeve type
Pneumatic system	No. 1 electrocompressor, No. 1 50-litre tank, No. 20 solenoid valves.

4 EXPERIMENTATION

Already in 1990-91 - as was mentioned in paragraph 2 - A.M.I. had experimented using a rented EDR plant. The groundwork of data and experienced collected at the time was a

AZIENDA MULTISERVIZI INTERCOMUNALE IMOLA

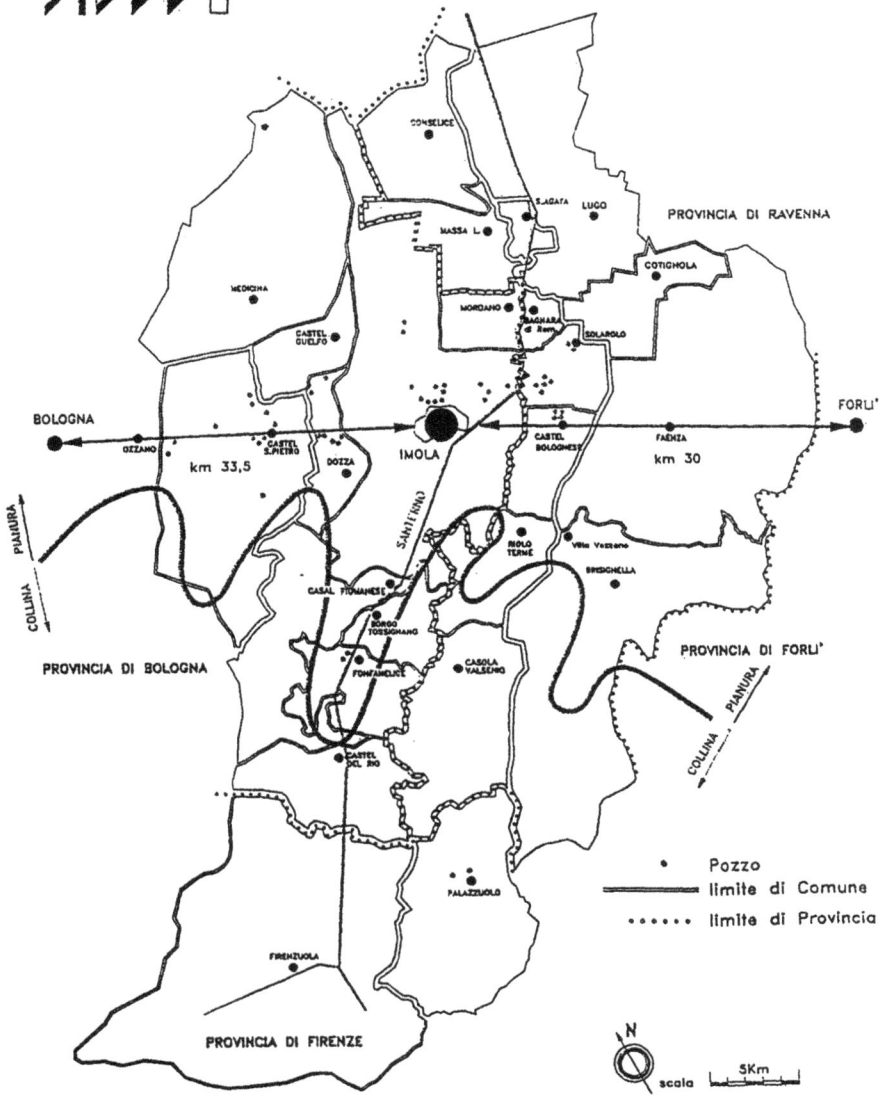

Figure 4

starting point for A.M.I. in developing the tender specifications for the supply of the current pilot plant (see block diagram of the system). We would like to underline here the creative role of A.M.I. in establishing "minimum system parameters" which the manufacturer had to comply with, and especially the direct involvement of the manufacturer in the following stage of performance of the experimental tests.

The programme of experimental tests was divided into two stages:

- The first stage was to check the achievement of the "minimum parameters" set down by A.M.I. in the plant supply specifications.
- During the second stage, after the steps decided on during the first stage were achieved, variations were set to the machine cycle in order to make improvements, both general and finalised to specific ends (reducing the consumption of additives, greater recovery, letting out less concentrated brine).

The following "minimum parameters" which it was necessary to comply with were set during the first experimentation stage:

a) Ion $NO3^-$ removal percentage: 70% (as an average)
b) specific power consumption: 0.79 kWh/m3 water production
c) percentage of brine discharge: < 15%

Simultaneous compliance with the three established parameters also had to be guaranteed for minimum continuous period of time, whatever variations there might be in the quality of the incoming water. Figure 5 is a graph showing the incoming and outgoing trend (from the treatment) of ion $NO3^-$ concentration, and the relevant abatement percentages. Fig. 6 shows the trend of the technical and management data gathered during experimentation (kWh/m3 ratio and relevant recovery percentage). The data show that the conditions set at the beginning were not only complied with; they were even improved upon (especially in the case of parameter b). Parameter c) is a special matter. Although this value can be set easily on the system, operating on an incoming by-pass valve, its achievement when associated with simultaneous achievement of the other two parameters is not so automatic, especially in the case of long term tests.

We can summarise as follows the plant conditions that were set and kept throughout the first phase (April to November 1997):

- polarity reversal cycles every 55 minutes
- acid washing with HCl 2% solution every 23 production hours
- antiscaling product proportioning: about 2 ppm on the concentrated flow
- sulphamic acid proportioning on the electrolyte circuit, depending on conductivity set up value of 15000 μS/cm.

As one can see, the tests during the first experimentation phase displayed two outstanding features:

a) high percentages of water recovery: >90%
b) high frequency of acid washing of the membrane packet (once a day, during the washing phase, since the plant had one line only, no treated water was produced).

The high percentage of water recovery, calculated as the ratio between water produced with a low nitrate concentration and the untreated water being fed in, obviously lead to the production of a concentrated discharge with high percentage of ion $NO3^-$.

Therefore, during the second experimentation phase, substantial modifications were made to the plant conditions in order to cut consumption down:

- polarity reversal cycles every 20 minutes
- acid washing using 2% HCl solution once a week
- sulphamic acid proportioning with conductivity set on the electrolyte at 10500 μS/cm.

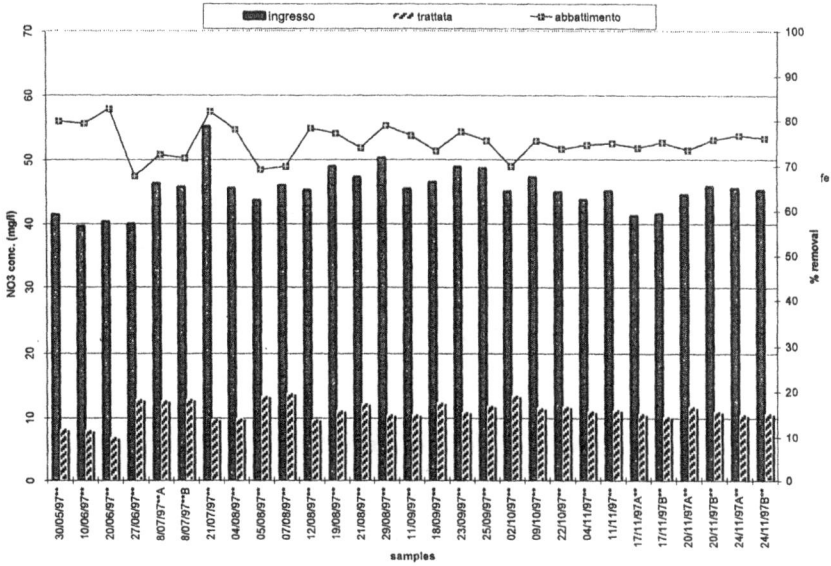

Figure 5 *NO₃ feed water conc. vs. treated water conc. and NO₃ removal %*

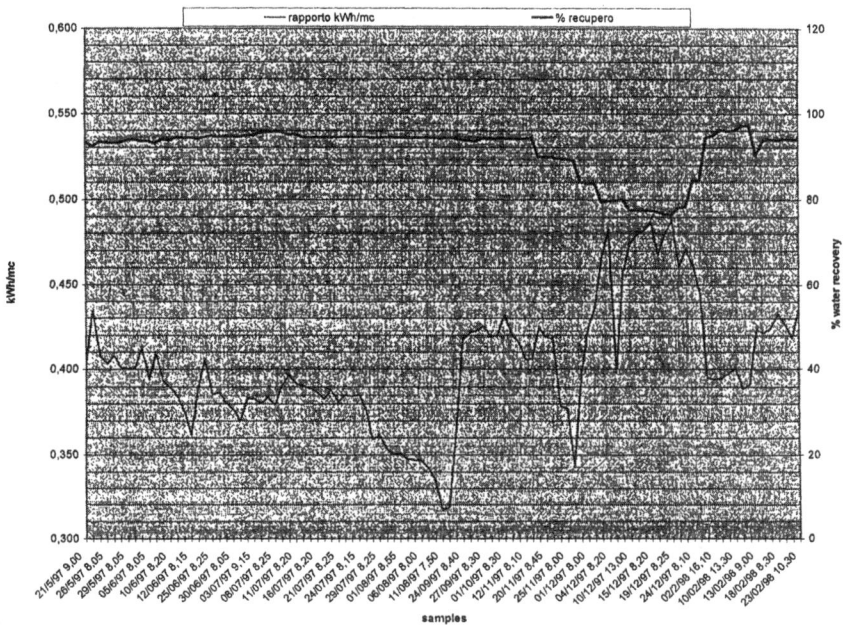

Figure 6 *kWh/me and % water recovery*

The application of these conditions - especially the reduction of the number of acid washings - led to a considerable reduction in the consumption of additives, while keeping the abatement throughput.

4.1 The quest for compatibility of the concentrated discharge.

The tests aimed at obtaining a discharge compatible with the limits set out earlier were run working progressively on the reduction of the percentage of water recovery: practically speaking. It was decided to increase the quantity of water drawn in by-pass from the input and delivered to the brine tank to dilute it; this led to an increase of the quantity of discharged brine.

This operation was performed in stages, lowering the yield from 94% step by step to 77%.

Measurements of the concentration of NO_3^- ion for discharge showed the following general trend:

- with recovery around 94%, there is a concentrate discharge between 500 and 600 mg/l.
- with recovery around 90%, there is a concentrate discharge between 300 and 330 mg/l.
- with recovery around 80%, there is a concentrate discharge between 200 and 250 mg/l.

On the basis of incoming concentrations of untreated water of around 50 mg/l of NO_3^- ion, one can make a rather accurate estimate on the basis of which one can say that in order to reach the limit set previously. It will be necessary to achieve recovery values around 70%.

The important factor which emerged from the tests that were made is that - despite the concentration of recirculating brine having been heavily reduced (from around 7000 μS/cm to around 2500 μS/cm), the percentage of ion nitrate abatement on the water produced is always near the previous levels, i.e. 70%. Therefore, the decision to preserve high percentages of water recovery was dictated by reasons of opportunity (especially in financial and management terms) and not by any technical limits of the EDR technology.

4.2 Assessing management costs

On the basis of the technical and management data gathered during the experimentation, we can draw up a sufficiently accurate estimate of management costs on the basis of the size of the plant. The following data cannot be applied directly to plants with a larger throughput, but it is quite probable that costs will diminish, since some will be spread out over larger production (additive consumption, labour, and analyses).

Total operating costs, on the basis of the experimental data, are lower than the 337 it£/m3(Italian Lire) of the data submitted by the Milan CAP which refer to a Reverse Osmosis treatment system, for a similar application[3].

Pilot Plan Throughput m3/year	About 205,000 m3 (average NO_3^- ion concentration of less than 15 mg/l)
Electric Power it£/m3 (estimate 250 It/kWh)	100
Running Consumption (additives) it£/m3	7
Routine Maintenance it£/m3	103
Personnel it£/m3	38
Chemical Analysis it£/m3	36
TOTAL OPERATING COSTS it£/m3	**284**

5 CONCLUSIONS

We can definitely say that the results of the experimentation made by A.M.I. together with the builders of the plant (FRAME S.p.A. of Ozzano Emilia, in the province of Bologna, Italy) are positive.

The data gathered on about 5000 hours of operation of the plant allow us to draw several important conclusions:

- during the last 8 years (the previous experimentation was in 1990), EDR technology for the abatement of nitrates from underground water has met with interesting developments, especially in terms of the performance of the membranes, and this leads us to expect further future developments;

- EDR technology has turned out to be very well focused as an application for treating the removal of nitrates in underground water, especially in terms of preservation of the taste and mineral balance features of the original water;

- EDR technology has absolutely competitive features in terms of management and investment costs when compared to high-tech processes such as Reverse Osmosis. Compared to other technologies also applied in treatment of the removal of nitrates from underground water (Ion Exchange and Biological Denitrification), EDR definitely involves higher investment costs, but still offers an interesting margin of competitiveness in terms of management costs, especially if we consider how much more flexible it is than a biological system (difficulty in controlling the process and critical plant start and stop phases).

- EDR technology, together with Reverse Osmosis (technologies with a higher degree of innovation) certainly present the best features in terms of flexibility and adaptability (at least for the specific application referred to here), especially when we consider that it was already possible during the planning stage to formulate tailor-made solutions, depending on the results which were desired; also, once the plant has been built, it can be easily adapted to new requirements without any special complications in terms of planning and/or management;

- Application of EDR technology for the treatment of underground drinking water presents no special problems in terms of adaptation, since this technology has already been used widely and for a long time in the food industry. Indeed, in the water distribution field conditions are far less critical than in other applications (concentration of the fluid being treated, temperature and pH conditions, etc.), so the life of the main components (membranes and electrodes) is certainly longer. Therefore, the amortisation costs of the plant are less.

References

1 P. Cotè: INTERNATIONAL REPORT "State of the art techniques in reverse osmosis, nanofiltration and electrodialysis in drinking-water supply - Anjou Recherche, Centre de Recherche de la Compagnie Generale des Eaux, Chemin de la Digne, BP76, 78603 Maisons Lafitte Cedex, France.
2 American Water Works Association: Manual of water supply practices - Electrodialysis and Electrodialysis Reversal - M38 (1995).
3 G. Gariboldi, S. Moriggi: "Esperienze applicative di trattamento dei nitrati - trattamenti con Osmosi Inversa" - XI Corso Residenziale sull'acqua, CISPEL Lombardia, Milan, May 21st-22nd, 1996.

NANOFILTRATION FOR DRINKING WATER TREATMENT FROM A EUTROPHIED LAKE IN TAIWAN

Hsuan-Hsien Yeh, Sheng-Herng Lin, Shan-Jhen Kao, and Grace T. Wang

Department of Environmental Engineering,
National Cheng Kung University,
Tainan 70101 Taiwan

1 INTRODUCTION

Cheng Ching Lake Water Works, located in southern Taiwan, is the main supplier of domestic water for the Greater Kaohsiung Area, the second largest metropolis in Taiwan with a population of over two millions and the location of major heavy industries. The water works drew its raw water from a nearby lake, a man-made off-line reservoir storing surface water pumped from the Kao-pin River. Owing to upstream discharge of farming, industrial, and domestic wastes, the water of the Kao-pin River is polluted. Therefore, the lake also becomes eutrophied. Although the treated water from the water works, which employs conventional treatment processes, including prechlorination, coagulation, sedimentation, filtration, and disinfection, can meet the current water quality regulation, the taste and odour causing compounds from algae can not be completely removed, and the hardness of the water is high. Therefore, the consumers have lots of complaints about the water quality.

Recently, the use of membrane technology in water treatment has developed quite rapidly. Membrane processes have the advantage of occupying less landspace and are flexible in expansion. Nanofiltration (NF), one application of the membrane technology, is reported being able to reject higher than 90% of the divalent and trivalent ions[1]. However, its rejection for the monovalent ions is only 30~60%. NF was first successfully used for ground water softening in Florida, USA[2]. Furthermore, it also has been employed for the removal of dissolved organics, including the precursors of halogenated organics. Its rejection for THMFP and AOXFP can be higher than 90%[3].

In order to upgrade the water treatment processes, a pilot scale testing, including nanofiltration, has being conducted to treat the raw water from the Cheng Ching Lake for simultaneous organic and hardness removal. In addition to water quality monitoring, the efficacy of various pretreatment strategies used to prevent membrane fouling was also studied.

2 MATERIALS AND METHODS

2.1 NF Membrane Pilot System

The core of the NF membrane pilot system is a two-stage membrane unit. The unit was configured with two pressure vessels; in the first stage feeding the concentrate to a single pressure vessel in the second stage. The permeate from each stage was collected and combined into a single product stream. Each pressure vessel is consisted of three elements in series, with a total of nine elements. The element used in this study was the spiral wound NF70 thin-film composite membrane (FilmTec, Dow Chemical).

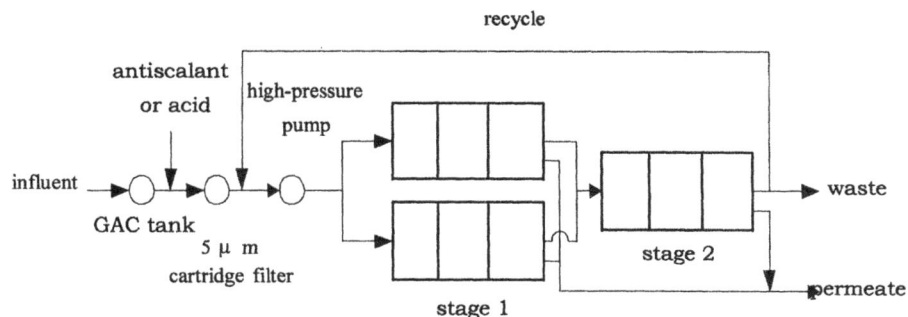

Figure 1 *Schematic Diagram of the NF Membrane Pilot System*

Each element was 10.2 cm (4 inch) in diameter by 101.6 cm (40 inch) long, with 7.62 m^2 (82 ft^2) of active membrane area. In order to meet the membrane specification of minimum flow, recycle stream was routed from the final concentrate stream back to the suction side of the booster pump blending the recycle with the membrane feed stream.

The NF membrane pilot system also includes an influent pump, a GAC tank, dosing pumps for acid or antiscalant, a 5μm cartridge filter, and a high-pressure feeding pump (Figure 1).

2.2 Pre-treatment Scheme

Before the water was fed into the NF system, the raw water from the Cheng Change Lake was subjected to various pretreatments. The purpose is to compare the efficacy of these pretreatments on membrane fouling control. The pre-treatment schemes tested in this study included: (1) coagulation, sedimentation, and granular media filtration (called conventional treatment process, CTP); (2) preozonation and CTP; (3) CTP and UF; (4) preozonation, CTP and UF.

For the CTP, conventional paddle type mixing tanks were used for both rapid and slow mixing. The coagulant used was liquid aluminium sulphate (7.5% Al_2O_3), with a dosage between 50~100 mg/L. Tube settler was utilised for sedimentation with an overflow rate varying from 18.8 to 69.8 m^3/m^2·day. Dual media filters, with 35 cm depth of anthracite (E.S. 0.89 mm) laying over 25 cm of silica sand (E.S. 0.51 mm), were used to further treat the settled water. The filtration rate was between 100 to 200 m^3/m^2/day. The goal of operating the CTP, either with or without preozonation, was to control the turbidity of the effluent from the granular media filter to be lower than 0.3 NTU over 95% of the time.

The UF membrane unit (Microza UF system, Asahi Chemical Industry, Japan) was consisted of two hollow fibre membrane modules (LGV-5210), made of polyacrylonitril (PAN), in parallel. The length and diameter of the module were 2227 mm and 140 mm, respectively, while the inner and outer diameters of the fibre were 0.8 and 1.4 mm, respectively. The effective surface area of each module was 41 m^2. The molecular weight cut-off (MWCO) of the membrane was 13,000. The filtration mode was outside to inside.

2.3   The Evaluation of NF Membrane System Performance

In order to control inorganic salt scaling, the pre-treated raw water was either subjected to pH adjustment to about 6.2 by adding sulphuric acid (or hydrochloric acid) or antiscalant application (Hypersperse™ AF200UL, Argo Scientific) before flowing through the 5 m cartridge filter and finally to the NF module.

The evaluation of the NF performance under various pretreatments is based on the decline of water mass transfer coefficient (MTC$_w$) with time. MTC$_w$ is determined by dividing the water flux of each membrane element by the net driving pressure:

$$MTC_w = \frac{F_w}{NDP} \qquad \text{(Eq.1)}$$

where MTC$_w$ = Water mass transfer coefficient (LMH/kg/cm^2)

    F$_w$ = Water flux of each membrane element (LMH)

    NDP   = Net driving pressure (kg/cm^2)

NDP is defined as:

$$NDP = \frac{P_I + P_C}{2} - P_p - \Delta\Pi \qquad \text{(Eq.2)}$$

where   P$_I$ = Inlet pressure

        P$_C$ = Concentrate pressure

        P$_P$ = Permeate pressure

        $\Delta\Pi$ = Net osmotic pressure

$$= (\frac{TDS_I + TDS_C}{2} - TDS_P) \times 0.0703 \,(\frac{kg/cm^2}{100mg/L})$$

TDS$_I$, TDS$_C$ and TDS$_P$ are the total dissolved solid concentrations of inlet, concentrate and permeate, respectively.

Once MTC$_w$ for each set of monitored pressure and water flux is determined, the MTC$_w$ for the entire operation period is plotted as a function of operation time (or cumulative permeate volume). The slope of the graph can then be obtained using statistical regression techniques. The slope is called the fouling rate of the membrane. When MTC$_w$ decreases to 85% of its original value, the membrane needs to be cleaned. The interval between cleaning is called the cleaning frequency (CF), noted by days.

During this testing, the influent, permeate and concentrate of the UF and NF systems were collected periodically for water quality analysis. The parameters analysed included turbidity, alkalinity, hardness, total dissolved solids (TDS), conductivity, UV absorbance at 254 nm (A254) and nonpurgeable dissolved organic carbon (NPDOC).

3 RESULTS AND DISCUSSION

Sixteen NF runs were carried out during the period from January 1999 to September 1999. Table 1 lists the operation conditions of all these runs. The recovery for the whole system was set at 75%, except Run 12, which was 60%. Figure 2 shows the plot of the MTC_w values obtained from Stage 1, Stage 2, and the whole NF system, respectively, versus the cumulative permeate volume for all these runs. The fouling rate and CF for each run are listed in Table 2.

In the first 9 runs, no matter whether the pre-treatment was CTP with or without preozonation, the CF of NF system was always less than 7 days, with relatively high silt density index (SDI) of the influent (about 5.2). At that point, it was assumed that the influent to NF system contained a high level of colloid and/or particulates thus fouled the NF membrane quickly. As a matter of fact, some of the runs were terminated due to clogging of the 5μm cartridge filter by algae, causing rapid headloss accumulation. This also indicated that some algal cells passed through the granular media filter, despite the turbidity of the filtrate had been controlled to be less than 0.3 NTU over 95% of the time.

Based on the assumption stated above, the UF unit was introduced to the membrane system from Run 10 (March 29, 1999). The filtrate from the granular media filter was further treated by UF before it entered the NF unit. As shown in Table 3, UF proved to be very effective for colloid and particulates removal, as SDI decreased from 5.2 to 0.8, and turbidity from 0.21 to 0.03 NTU. There was about 17% removal of NPDOC, probably by adsorption on the fibre surface. However, removal on dissolved species, such as TDS, conductivity, and hardness, was not observed. The CF for Run 10 was still very short (3 days only, as shown in Table 2). The over-all efficacy on NF membrane fouling control by applying UF as pre-treatment was minor. This suggested the colloid and particulates probably were not the primary factors causing membrane fouling.

Another possible factor, the soluble species, was then taken into consideration. To investigate the major composition of the foulant, the spent cleaning solutions were analysed. In the first four runs, the cleaning agent used was hydrogen peroxide solution. However, after the cleaning, the original MTC_w value (7.48 LMH/kg/cm^2) could not be re-established. The cleaning efficiency of hydrogen peroxide solution was doubtful. From Run 5 to 9, the NF membrane was first washed by sulphuric acid solution (pH=2) and then followed by 45% sodium hydroxide solution wash. It can be found in Figure 2 that the MTC_w value after the acid-base wash resumed its original value. In Run 10, only acid wash was employed for membrane cleaning and the restoration of MTC_w was just the same as that of acid-base wash used in Runs 5 to 9. This probably indicates that inorganic scales mainly caused the fouling.

From Run 1 to 10, the depression of pH level of the influent was achieved by adding H_2SO_4. However, the concern of adding H_2SO_4 was that the fouling may be caused by the deposition of sparingly soluble sulphates salts, such as $CaSO_4$, $SrSO_4$, or $BaSO_4$. Therefore, since Run 11, the acid employed was switched from H_2SO_4 to HCl. Nevertheless, the improvement in fouling control was minor, since the CF only increased to 5.7 days. Later, when the water quality analysis data of the feed and the acid-wash waste were available (Table 4), it was surprising to find a tremendous increase in aluminium concentration in the acid wash waste, as compared to that in the influent. This suggested that Al might be the major cause for the membrane fouling.

Table 1 *Operation Conditions of the NF System**

Run	Date	Feed Pressure (kg/cm^2)	Recovery (%)	Pre-treatment**	Acid for pH Control	Cleaning Agent
1	11/1/99~15/1/99	5.48~7.25	75	1	H_2SO_4	H_2O_2
2	15/1/99~18/1/99	5.65~6.9	75	2	H_2SO_4	H_2O_2
3	18/1/99~22/1/99	5.5~6.6	75	2	H_2SO_4	H_2O_2
4	3/2/99~6/2/99	5.65~6.97	75	1	H_2SO_4	H_2O_2
5	9/2/99~13/2/99	5.3~7	75	2	H_2SO_4	$H_2SO_4 + NaOH$
6	9/3/99~11/3/99	4.66~7.3	75	1	H_2SO_4	$H_2SO_4 + NaOH$
7	16/3/99~19/3/99	5.7~6.61	75	1	H_2SO_4	$H_2SO_4 + NaOH$
8	22/3/99~25/3/99	4.95~6.6	75	1	H_2SO_4	$H_2SO_4 + NaOH$
9	25/3/99~29/3/99	4.67~7.1	75	1	H_2SO_4	$H_2SO_4 + NaOH$
10	29/3/99~2/4/99	4.65~7.05	75	3	H_2SO_4	H_2SO_4
11	12/4/99~19/4/99	4.95~6.95	75	4	HCl	H_2SO_4
12	21/4/99~27/4/99	4.15~5.5	60	5	HCl	H_2SO_4
13	20/5/99~29/5/99	4.15~5.5	75	5	HCl	H_2SO_4
14	21/6/99~3/7/99	4.8~6.68	75	5	HCl	H_2SO_4
15	3/7/99~16/7/99	4.8~5.85	75	5	H_2SO_4	Bioclean 103A + Bioclean 511
16	4/8/99~14/9/99 (w/ antiscalant)	4.8~5.3	75	6	none	

* pH of the membrane feed stream was controlled at 6.2 for all runs, except Run 16.

** Pre-treatment Schemes:

1. CTP (Conventional treatment processes) + 5 µm cartridge filter
2. Preozonation + CTP + 5 µm cartridge filter
3. CTP + UF
4. Preozonation + CTP + UF
5. CTP + UF + 5 µm cartridge filter + coagulation pH control at 6.8
6. CTP+UF+5µm cartridge filter

Figure 2 *Plots of MTCw versus Cumulative Permeate Volume*

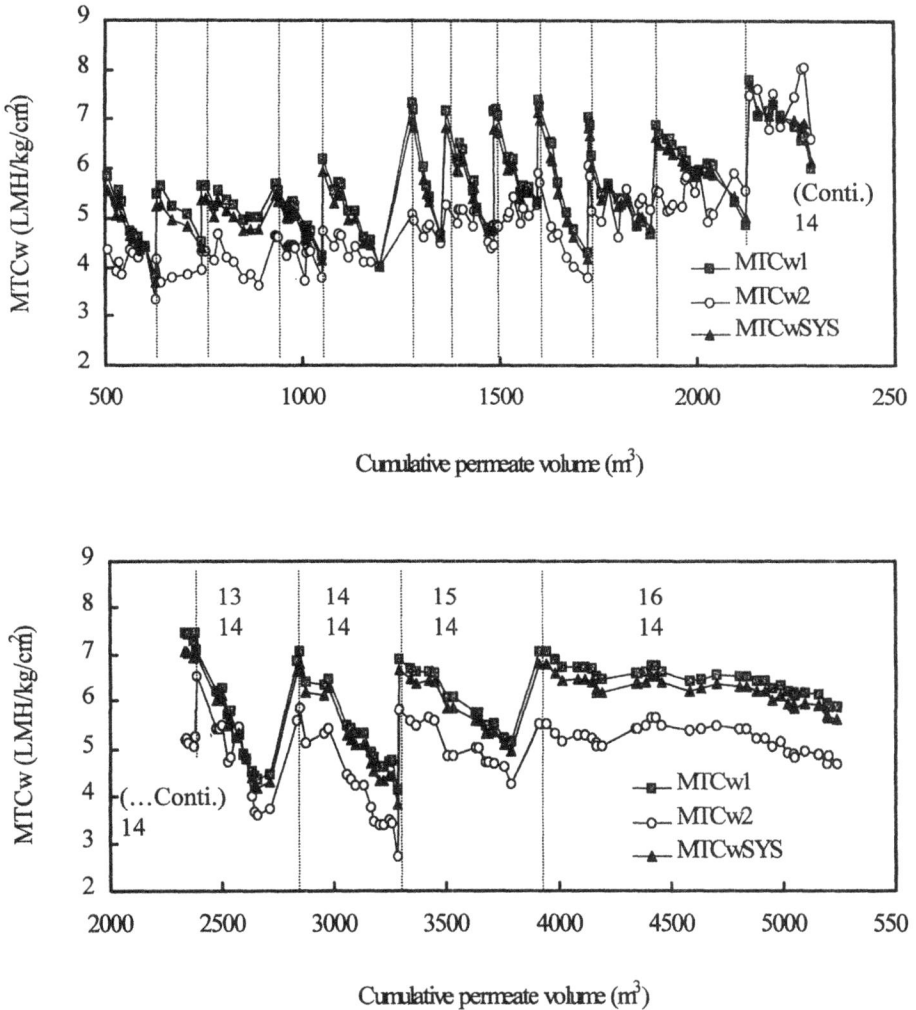

Cumulative permeate volume (m³)

In order to allow the least amount of Al into the membrane system, the operation of rapid mixing tank of CTP was controlled at pH 6.8 by adding hydrochloric acid in Runs 13 and 14, since the solubility of Al salt is least at this pH level[4]. Although these 2 runs seemed to last longer than the previous ones, the improvement was limited. In the following run (Run 15), sulphuric acid was once again used to control the pH level of coagulation in CTP and the influent to the NF system. It was found that the performance of the NF system did not deteriorate. Therefore, at this stage, it could not be determined whether the fouling of NF was caused by sulphate precipitation.

Table 2   *Fouling Rate and Cleaning Frequency (CF) of NF Operation*

Run	Fouling Rate (LMH/kg/cm^2)	CF (day)	
1	0.475	2.4	
2	0.254	4.5	
3	0.166	6.9	
4	0.363	3.1	
5	0.441	2.6	
6	1.176	1.0	
7	0.603	1.9	
8	0.511	2.2	
9	0.858	1.3	
10	0.377	3.0	
11	0.200	5.7	
12	0.227	5.0	
13	0.306	3.7	
14	0.209	5.4	
15	0.119	9.4	
16	0.021		53.4

$MTCw_i = 7.48\ LMH/kg/cm^2$

Table 3   *Water Quality of UF System*

	eed stream	ermeate	emoval (%)
Turbidity (NTU)	.21	.03	86
Hardness (mg/L as CaCO$_3$)	57	51	2.3
Alkalinity (mg/L as CaCO$_3$)	54	56	1.3
NPDOC (mg/L)	.89	.74	6.9
TDS (mg/L)	72	73	0.3
Conductivity(µS/cm)	42	51	1.7
SDI	.2	.8	85

Before starting Run 16, the NF membrane was thoroughly cleaned by two proprietary formulated cleaners. The membrane was first washed with a low pH liquid formulation (Bioclean 103A, Argo Scientific) for one hour, then followed by a high pH liquid formulation (Bioclean 511, Argo Scientific) for another hour. Based on the pressure required to maintain certain flux after each cleaning as well as the water quality analysis (*e.g.*, Al, Si, Ca, *etc.*) on the cleaning solution waste, it can be noticed that acid cleaning was more effective than alkaline cleaning. The performance of these cleaning products is superior to that of the H$_2$SO$_4$ /NaOH solutions used in previous runs.

In Run 16, a phosphonate and polymer based antiscalant (Hypersperse™ AF200UL, Argo Scientific) was added continuously into the influent of the NF system at a dosage of 3 mg/L. The pH of coagulation, prior to the 5µm cartridge filter, by addition of acid was suspended. The result was very promising and the CF was extended to about 53 days. So far, the application of antiscalant provided the best NF performance.

Table 5 lists the average water quality of the influent and permeate of the NF system. More than 80% of total hardness, TDS and conductivity were removed. The rejection for organic parameters, such as NPDOC and A254, is 75 and 72%, respectively. The organoleptic quality of the NF permeate was also superior to that of the tap water from Cheng Ching Lake Water Works, according to blind tests by consumer groups. One of the advantages of the membrane process is its stable permeate quality. Based on the result from these 16 runs, the permeate quality was consistent over the operation period.

Table 4 *Water Quality Analysis of the Feed and Acid-wash Waste obtained from Run 10*

	Ion Concentration (mg/L)			
	Ca^{2+}	Ba^{2+}	Sr^{2+}	Al^{3+}
Feed	78	0.0194	0.517	·0.122
Acid-wash waste	83.2	0.0182	1.107	52.56

Table 5 *Water Quality of NF System*

	Feed stream	Permeate	Removal (%)
Turbidity (NTU)	0.06	0.03	50
Hardness (mg/L as $CaCO_3$)	233	20	91
Alkalinity (mg/L as $CaCO_3$)	121	22	82
A254 (m^{-1})	1.63	0.45	72
NPDOC (mg/L)	0.675	0.17	75
TDS (mg/L)	247	47	81
Conductivity($\mu S/cm$)	521	105	80

4 SUMMARY

This pilot scale testing, including NF membrane unit has led to a successful upgrade of the treatment processes of a water works which drew its raw water from an eutrophied lake. The NF membrane process was found to produce high quality permeate, with over 75% of NPDOC rejection, 90% of hardness rejection, and excellent organoleptic quality. However, pre-treatment should be carefully chosen to prevent fouling. Several pre-treatment schemes were studied, including conventional treatment process only, conventional treatment process with preozonation, conventional treatment process with UF, and antiscalant application. Although conventional treatment process with or without preozonation could reduce turbidity to less than 0.3 NTU, the SDI was still high. When UF was added, the SDI reduced to 0.8, indicating that most colloidal and particulate foulants were removed. However, scaling caused by inorganic precipitation still can not be solved. So far, the application of antiscalant provided the highest efficacy on NF fouling control.

5 ACKNOWLEDGEMENTS

The financial supports provided by National Science Council, Taiwan, ROC (Project No. NSC 87-2211-E-006-010), Taiwan Water Supply Corporation, and China Steel Company are greatly appreciated. The assistance provided by Dow Chemical Company, Asahi Chemical Industry Co., and BetzDearborn Company during this study is also acknowledged.

References

1. AWWARF, LdE, and WRC, *Water Treatment Membrane Processes*, McGraw-Hill, New York, 1996.
2. W.J. Conlon and S.A. McClellan, *Jour. Am. Wat. Wks Asso.* 1989, **81**, 47.
3. T.J. Blau, J.S. Taylor, K.E. Morris, and L.A. Mulford, *Jour. Am. Wat. Wks Asso.* 1992, **84**, 104.
4. S. Kawamura, *Intergrated Design of Water Treatment Facilities,* John Wiley & Sons, New York, 1991.

Fouling and Cleaning

MEMBRANES AND MICROORGANISMS - LOVE AT FIRST SIGHT AND THE CONSEQUENCES

H.-C. Flemming

Department of Aquatic Microbiology
University of Duisburg
Geibelstraße 41
D-47057 Duisburg, Germany

1 INTRODUCTION

The treatment of water by membrane technology intrinsically implies the contact of very large quantities of water with the membrane surfaces. This water is not sterile. In drinking water, the numbers of cells actually present as demonstrated by microscopic quantification usually range between 10^4 and 10^6 cells mL^{-1}. These cells have a tendency to adhere to surfaces; in oligotrophic systems, this is considered a survival strategy[1]. There is virtually no surface material which cannot be colonised, even under extreme conditions[2], regardless of hydrophobicity or hydrophilicity, smoothness or chemical composition[3] - surface conditions and materials will simply select for colonising species among the spectrum of organisms in a given water volume. In a membrane system, adhesion to the membrane surface is facilitated by the vertical transport vector which is given by the water flow through the membrane[4] - this can be described metaphorically as "love at first sight", because there will always be some organisms which prefer to settle on the given membrane material, be it hydrophobic or hydrophilic. Once the organisms colonise the surface, they will inevitably multiply and form biofilms. All membrane systems, which are not operated under absolutely sterile conditions, will carry biofilms[5].

Not all of the systems carrying biofilms suffer from biofouling - "biofouling" is an operational term, applied when the effects of biofilms exceed a certain threshold, or tolerance level, which is individually set for different systems[6].

In membrane systems, however, biofouling is the "Achilles heel" of the process, because all other fouling components, such as organic and inorganic dissolved substances and particles can mostly be removed by efficient pre-treatment; however, microorganisms are particles which can multiply. Thus, if they are removed to 99.99 %, there are still enough cells left which will grow at the expense of biodegradable substances in the water. Microorganisms are ubiquitous in any technical system unless it is kept sterile by enormous and continuous effort. The types of microorganisms and the nutrient concentration represent the biofouling potential.

Biofouling leads to considerable technical problems and economic loss[7]. Not only the feed side of the membranes is concerned but microorganisms can also pass the membrane[8], although they are believed to be too large to penetrate a reverse osmosis membrane. As membrane technology will be of increasing importance in meeting the rising demand for treated water, biofouling will be an increasing problem. Biofouling is a biofilm problem; it is important to understand basic biofilm processes and properties in order to design rational countermeasures.

2 ADHESION AND COHESION

Within minutes of contact with non-sterile water, the first microorganisms will adhere[4]. Primary colonisation is strongly influenced by the concentration of cells in the water phase. Among the spectrum of various microbial species, a clear preference for given membrane materials can be observed; one example is the preferential colonisation of polysulfone by *Pseudomonas diminuta* and *Staphylococcus warneri* as investigated by Flemming and Schaule[10]. In order to investigate such preferences, samples of the mixed population of a mature biofouling layer were selectively removed from the bottom (cell-membrane interface) and the top of the layer. Various membrane materials were exposed to suspensions of the two populations and the adhesion kinetics were measured. While the bottom population showed a clear preference for the material from which it was isolated, the top population did not. This makes sense as the population at the water-biofilm interface is separated from the membrane material by many layers of microorganisms and extracellular polymeric substances (EPS). Thus, manipulation of the membrane material in order to reduce microbial adhesion most probably will select for a species that adheres to that material. With time, this species will cover the surface and mask its effect. Similar observations have been made with anti-fouling coatings on ship hulls, in particular, with copper plating.

It was shown that dead cells of *P. diminuta* adhere at the same rate to the surface as living cells[9]. This indicates that these cells already carry the "glue" in suspended form, and the material which mediates adhesion as well as cohesion is the EPS. These are composed of polysaccharides, proteins, glycoproteins, lipoproteins and other macromolecules of microbial origin[10]. They form a slime matrix, which sticks the cells to the surface and keeps the biofilm together. Any cleaning measure has to overcome the overall binding energy of this system. This energy is not provided by covalent chemical bonds, but by weak physio-chemical interactions. In general, they can be divided into electrostatic interactions, hydrogen bonds and van der Waals interactions (fig. 1). The average binding energy ranges between 0.1-10 % of that of a covalent C-C bond, depending on the respective conformations of the macromolecules, the water content, pH value, ionic strength, temperature and other parameters. The weak binding energy of the individual bonds is increased by the fact that the EPS molecules possess many functional groups capable of interaction. If a macromolecule has 10^6 possible binding sites and only 10 % of them are interacting; the binding forces of the weak interactions are multiplied by a factor of 10^5, resulting in a considerable stability. The mechanism by which cleaners disintegrate biofilms is based on interference with these interactions. They do contribute, but not always in the same proportion, to the overall binding forces - they vary, according to surface properties and EPS composition. Cleaners have to overcome the cohesion forces and have to address all forms of weak physio-chemical interactions.

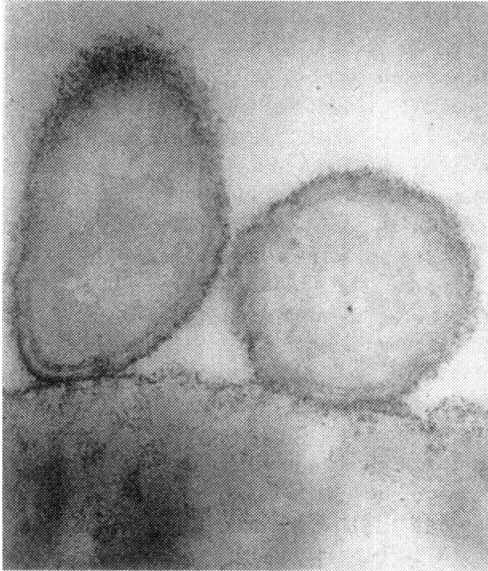

Fig. 1: *Primary adhesion of P. diminuta to polysulfone membrane material*

Surfactants are a major constituent in many cleaning formulations. They will interfere with van der Waals interactions and influence the so-called hydrophobic interactions. Van der Waals interactions can be dominant in systems in which cells adhere from water to hydrophobic surfaces. Schaule[11] has shown that the adhesion of *P. diminuta* to polysulfone membranes is performed with significant participation of van der Waals interactions. This could be demonstrated by the influence of surfactants on the adhesion rate: however, in preliminary experiments, the cohesion of alginate, an extracellular polysaccharide of *Pseudomonas aeruginosa*, was not affected by surfactants, but was strongly influenced by electrostatic interactions[2]. These examples demonstrate that in different systems, different binding forces can dominate.

Phosphate, citric acid, salts, other ionic compounds and complex-formers will interfere with weak electrostatic interactions. Many of these substances are components of cleaners. It has been shown that electrostatic interactions, which are important in cohesion of EPS molecules12, do not prevail in adhesion of *P. diminuta* to hydrophobic polysulfone membrane surfaces[11].

So-called chaotropic agents, such as urea, tetramethyl urea, guanidine hydrochloride, and others which are known from protein chemistry interfere with hydrogen bonds. They literally cause a chaos in water structure by rapidly binding water molecules, which are ripped from hydration, water surrounding proteins or polysaccharides. Hydrogen bonds represented a dominant kind of force in both adhesion and cohesion systems as described above; however, chaotropic agents are usually not constituents of cleaners. In cohesion and adhesion experiments using alginate as an EPS model, guanidine hydrochloride showed significant effects[11,12]

The entanglement of the macromolecules provides an additional factor in biofilm stability. This is addressed by the use of enzymes. The problem, however, is that the EPS macromolecules are of highly variable composition and structure, and enzymes mostly are too specific to act on the entire variety. This is one reason why enzyme treatment frequently yields disappointing results in practice.

In membrane systems, it has been shown that cleaners can improve the permeability of the fouling layer without reducing the fouling layer. In an experiment, an agar gel layer was taken as a model for a hydrogel[13]. Even though the layer thickness is not changed by the application of the cleaner, an almost five-fold increase in the layer permeability was seen. Thus, cleaners can improve (and decrease) the permeability of the biofilm, although this is a transient effect. The results suggest that optimisation of the fouling layer is possible even if it is not removed. This, however, is not true for every given cleaner. Correct selection and tailoring of conditioning agents has been shown to significantly alter the hydraulic resistance of biofilms. Increasing the permeability is desirable (i.e. decreasing the resistance or specific resistance). Table 1 shows some examples of changes in fouling layer permeability demonstrated with model fouling layers of filter cakes from bacteria and activated sludge. In the case of formaldehyde, a commonly used disinfectant, the effects of its application is to reduce the permeability of microbial layers substantially; that is not very surprising, as formaldehyde is used as a fixation agent in microscopy. This is an experimental example with artificial model biofilms, which helps understand why the performance can deteriorate after the use of cleaners.

Table 1.   *Alterations in specific hydraulic resistance, η and permeability, L_p of microbial layers due to chemical conditioning (after [13])*

Layer	Agent	$\dfrac{\eta_{after}}{\eta_{before}}$	$\dfrac{L_{p\,after}}{L_{p\,before}}$
Filter cake of bacteria	Commercial Cleaner (Anionic surfactant-non-ionic surfactant mix)	0.4	2.5
Filter cake Of bacteria	Tannin (1 % in water)	1.3	0.8
Activated sludge	Formaldehyde (1 % in water)	1.7	0.6

3 DETECTION

In fouling cases, it is important to distinguish different foulants in order to design effective countermeasures. Usually, the membrane system will respond to biofouling problems by an increase of $\Delta p_{feed/brine}$ and/or $\Delta p_{membrane}$, however, this response is very non-specific and can be caused by other fouling mechanisms as well. The diagnosis "biofouling" will be given if countermeasures, which are usually taken against non-biological foulants, fail. Then, a microbiologist is called and will most probably take water samples, count colony forming units or cell numbers, and maybe will identify a few of the microorganisms found in the water. In most cases, these data cannot be related to the location of biofilm growth, as the cells in the water can be released from any site in the system, including the membranes. As cells do not detach from biofilms at a continuous rate but randomly, water phase cell numbers cannot be related to the extent of biofilm growth either.

Thus, sampling on surfaces is mandatory. A good preliminary sampling protocol for ultrapure water systems is given by Sematech[14]. Unfortunately, there is still no technology available to take biofilm samples non-destructively from an operating membrane. Interesting approaches are under development, such as fibre optical sensors that can be integrated in a membrane module and report the development of deposits on surfaces by an increase of reflected light[15], but this sensor still has to be adapted to common membrane module designs.

In practice, either a bypass membrane device is used from which membranes can be removed and investigated destructively, or other accessible and representative surfaces such as cartridge filters are sampled. In order to evaluate biofouling problems fatal to a module, an autopsy is performed.

3.1.1. Standard procedure for module autopsy A standard procedure for module autopsy as carried out in our laboratory is the following:

1. The module arrives in the laboratory (preferably cooled, not preserved, not dried, and immediately after being removed from the plant), is opened mechanically and unfolded.

2. Optical inspection gives first information about the colour, the thickness and the consistency of the fouling layer. The colour can indicate the participation of humic substances and iron compounds in the material; it also can be caused by microbial pigments[16], which can be misinterpreted as inorganic foulants. Scratching on the surface gives an indication about the consistency of the layer.

3. A part of the membrane is inspected microscopically. This gives an indication about the prevalence of microorganisms and about the structure of the fouling film. In special cases, a cryosectioning procedure is carried out in which the membrane and the adjacent fouling layer are embedded in a water-soluble resin, which can be frozen and cut into slices of 2-50 μm. This procedure allows the exact determination of the thickness of the fouling layer[6].

4. A defined surface area is scraped off thoroughly. Water content and incineration loss are determined. A high water content and incineration loss are indicative for biofilms.

5. A part of the material is suspended in a defined volume of sterile water. The following parameters are determined in relation to the surface area:

- Cell number (by epifluorescence microscopy; it yields the maximum number of cells present in the sample).
- Number of actively respiring cells can be determined (by using a redox dye such as 5-cyano-2,3-ditolyltetrazolium chloride, CTC[17], which is used as an electron acceptor by the microorganisms).
- Number of colony forming units (on various nutrient media, which gives information about the minimum number of living cells. With regard to the particular fouling case history, the number of anaerobic or autotrophic bacteria, or other specific organisms, can be determined and related to the surface area.
- Content of proteins (determined after Lowry et al.[18])

- Content of polysaccharides (by a modification of the method of Dubois et al.[19])

- Content of ironic acids (indicative for acidic polysaccharides; determined after Blumenkrantz and Asboe-Hansen[20])
- Humic substances (after Frølund et al.[21])

3.1.2 FTIR-spectroscopic analysis: The use of FTIR spectroscopy in the analysis of microbial aggregates has been described in detail by Nichols et al.[22] and Schmitt and Flemming[23]. The attenuated total reflection (ATR) mode offers a further possibility to investigate smooth surfaces of various materials directly. This can be achieved practically without sample preparation. With respect to biofilm research, it offers the significant advantage that the sample can be investigated relatively undisturbed. Thus, a biofilm does not have to be removed from its support, which would alter its structure considerably from its native form. The ATR method allows one to measure in aqueous media and to investigate the development of a biofilm *in situ*, directly at the interface, non-destructively and in real time[23].

4 COUNTERMEASURES

The far most widespread approach against acute biofouling problems is the application of biocides. This originates from a medical kind of thinking: if microorganisms cause the problem, it is wise to kill them; however, this approach has some severe drawbacks in practice because it is applied to a technical system and not to a living organism. Therefore, a biocide may kill the biofilm organisms, although it is known that biofilm organisms display a significantly higher resistance to biocides[24], but it usually will not remove the biofouling layer. As the physical properties of this layer, i.e. the hydrodynamic resistance, is the problem, it does not help very much to kill the organisms and leave the biomass where it is (fig. 2). Cleaning should be understood as a two step process: (i) Weakening of the fouling layer (performed by cleaners which interfere with the dominating physio-chemical interactions), and (ii) Removal of the fouling layer, usually performed by shear forces. It is very important to assess the efficacy of the cleaning process, as it may be much less effective than expected. It is evident that the remnants of the fouling layer will invite new biofouling.

5 PREVENTION

The conventional anti-fouling strategy is to dose continuously with biocides[25]. Various membrane producers and vendors of water treatment provide dosing programmes systems; however, they do not always fulfil expectations, which is the reason for continuous research on additional and alternative strategies against membrane biofouling. Continuous biocide application generates wastewater problems. The cost of the treatment of this wastewater can exceed the savings gained by using the membrane technology.
    It has been demonstrated that practically every membrane system operating with water carries biofilms, unless it is operated under absolutely sterile conditions[5]. As a consequence, the biofilm participates in the separation process, although in an undefined

way; however, not every membrane system has biofouling problems. This indicates that biofouling is an operationally defined phenomenon - it only occurs if biofilm development exceeds a certain threshold of interference which may differ from one system to the other and is individually defined. Thus, a suitable anti-fouling strategy could be to tolerate biofilms below the threshold and curb the excess of biofilm growth above the threshold.

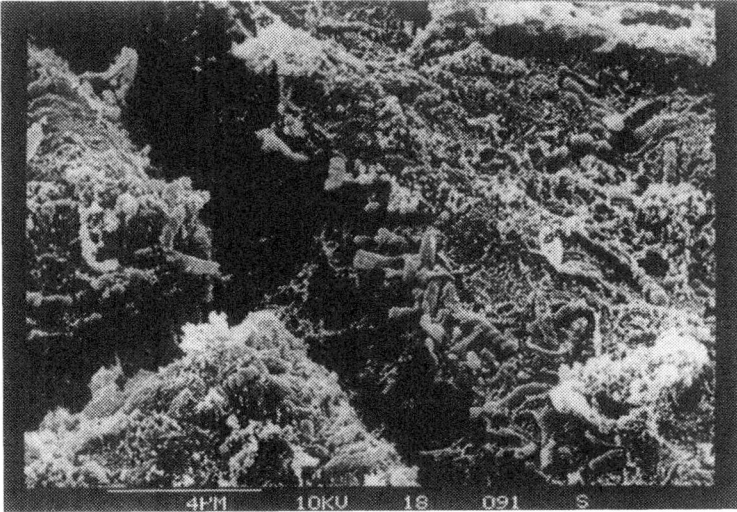

Fig. 2: *Scanning electron micrograph biomass left on fouled membrane surface after extended disinfection*

As indicated earlier, the biofouling potential comprises the ubiquitous microorganisms and the availability of nutrients. These must be considered as potential biomass. In technical systems, most biofilms will have reached the plateau phase of biofilm thickness. The extent of biofilm accumulation in this phase will depend on different factors that control the equilibrium level of the plateau:

*   Nutrient concentration, type and availability
*   Shear forces
*   Mechanical stability of the biofilm matrix, as influenced by
  - oxidising agents (biocides, etc.)
  - biodispersants
  - mechanical stress
  - temperature
  - type of microorganisms
  - physiological activity (gas production)
  - structure and physical strength of the EPS network
  - grazing organisms

The most important factors are nutrient concentrations and shear forces. A factor of minor importance seems to be the cell density in the water phase. In the plateau phase, the further adhesion of cells to the biofilm does not contribute significantly to biofilm accumulation[26]; however, these cells are targeted by an effective biocide treatment, while the nutrient concentration is not decreased. In some instances, the nutrient concentration may even be increased, when the biocide reacts with recalcitrant organic molecules, increasing their bioavailability[27].

Biofouling can be regarded as a biofilm reactor in the wrong place, as exactly the same natural processes occur in both situations. Thus, if a surface-rich area is offered ahead of a system, which should be protected against biofouling, biofilms will form there and consume degradable matter from the water stream. This will decrease the extent of biofilm development in subsequent compartments. Bott[26] investigated biofouling in heat exchangers and has clearly shown that decreasing the nutrient concentration could reduce the extent of biofilm accumulation. Biological filter systems such as those used in drinking water treatment might provide useful tools in reducing the nutrient concentration and thus, in prevention of biofouling. To date, biocides are worldwide the only answer to biofouling. Facing increasing difficulties in the application of biocides in both effectiveness and environmental regulations, membrane technology might be well advised to develop biocide-free anti-fouling strategies. The optimisation of nutrient limitation techniques could provide new possibilities. The role of assimilable organic carbon (AOC) will be very important in this approach. This approach would utilize biofilms in the right place, where they can be handled easily, in order to minimise the extent of biofilm formation at sites where biofilms need to be limited. Clearly, this concept cannot be applied to any given separation plant; however, if it is applied where appropriate, it might save large amounts of biocides. In addition, the concept needs reliable biofilm monitoring devices, i.e. representing biofilm accumulation on membrane surfaces. Sacrificial module elements, as already proposed[25] will be highly useful. Considerable research efforts will be required to put this concept into reality, but it might help to solve fouling problems in general. A practical example has been carried out in a water treatment system for a heat exchanger[6]. River water treated by flocculation and sedimentation was used as cooling water. A sand filter was integrated as a biofilter. Before and after the filter there was a reverse osmosis test cell installed, one operating with water before, the other with water after the biofilter (parameters given in table 2). Uronic acids were measured as indicative for EPS. The protein:uronic acid ratio before the sand filter was 7.7:1 (w/w) while it was after the sand filter 3.6:2.3. This indicates that the biofilm contains much more EPS under conditions of nutrient depletion. The BDOC is the biodegradable fraction of the dissolved organic carbon (DOC) in water. Interestingly, the thickness of the biofilm was reduced to a tenth of the original size, and the structure of the biofilm was different; it seemed that the biofilm formed under nutrient depletion was much less dense than the biofilm that formed in raw water. It was acceptable to reach a BDOC of 0.125 mg L^{-1} in order to stay below the threshold of interference. Such a value can be reached easily with common biofilters. The data on microbial concentration in the water phase show that the biofilter does not cause additional microbial contamination but rather removes cells from the process stream.

Table 2: *Biofilm, water and performance parameters before and after sandfilter (after [6])*

	Before sand filter	After sand filter
Biofilm		
Thickness	$(27.3 \pm 3,1)$ μm	$(3.0 \pm 0,5)$ μm
Protein content	77.7 μg cm^{-2}	3.6 μg cm^{-2}
Carbohydrate content	22.5 μg cm^{-2}	2.6 μg cm^{-2}
Uronic acid content	10.6 μg cm^{-2}	2.3 μg cm^{-2}
Water		
BDOC of water	0.325 mg L^{-1}	0.125 mg L^{-1}
Microbial content	1 x 10^7 cfu mL^{-1}	1.2 x 10^6 cfu mL^{-1}
Permeate production	65 % after 10 days	>95 % over test period

6 OUTLOOK

A rational anti-fouling strategy should reflect the properties and dynamics of biofilms. Fundamental aspects include sampling on surfaces, removal of the biomass, regarding nutrients as part of the fouling potential, and the integration of monitoring techniques for early recognition of fouling problems and for optimisation of countermeasures[15, 18]. Thus, biofouling cannot be overcome by application of some wonder-chemical but by a better understanding of the entire process and by interfering in the most effective way and on a scientific basis. This can include the "love at first sight" indicated in the title of this paper, because it is possible to live with biofilms as long as their effects remain below the threshold of interference. The combination of nutrient removal and monitoring represents a sustainable strategy for the future.

References

1. Marshall, K.C. (1996): Adhesion as a strategy for access to nutrients. In: Fletcher, M. M. (ed.): Bacterial adhesion. John Wiley, New York; 59-87
2. Flemming, H.-C. (1991): Biofilms as a particular form of microbial life. In: H.-C. Flemming and G.G. Geesey (eds.): Biofouling and Biocorrosion in Industrial Water Systems. Springer, Heidelberg; 3-9
3. Flemming, H.-C., Griebe, T. and Schaule, G. (1996): Anti-fouling strategies in technical systems - a short review. Water Sci. Technol., 517-524
4. Ridgway, H.F. (1988): Microbial adhesion and biofouling of reverse osmosis mem branes. In: Parekh, B.S. (ed.): Reverse osmosis technology. Marcel Dekker, New York, Basel; 429-481
5. Flemming, H.-C., G. Schaule and R. McDonogh (1993) How do performance parameters respond to initial biofouling on separation membranes? Vom Wasser 80, 177-186
6. Griebe, T. and Flemming, H.-C. (1998): Biocide-free antifouling strategy to protect RO membranes from biofouling. Desalination 118, 153-156
7. Characklis, W.G. (1990): Microbial fouling control. In: W.G. Characklis and K.C. Marshall (eds.): Biofilms. John Wiley, 585-633

8. Ghayeni, S.B.S., Beatson, P.J., Fane, A.J. and Schneider, R.P. (1998): Bacterial passage through microfiltration membranes in wastewater applications. J. Membr. Sci. 3972, 1-12

9. Flemming, H.-C. and G. Schaule (1988): Biofouling on membranes - a microbiological approach. Desalination 70, 95-119

10. Wingender, J., Neu, T. and Flemming, H.-C. (1999): What are bacterial extracellular polymeric substances? In: Wingender, J., Neu, T. and Flemming, H.-C. (eds.): Bacterial
    extracellular polymeric substances. Springer, Heidelberg, Berlin; 1-19

11. Schaule, G. (1992): Primäradhäsion von _Pseudomonas diminuta_ an Polysulfon-Membranen. Dissertation Univ. Tübingen

12. Mayer, C., Moritz, R., Kirschner, C., Borchard, W., Maibaum, R., Wingender, J., and Flemming, H.-C. (1999) The role of intermolecular interactions: studies on model systems for bacterial biofilms. Int. J. Biol. Macromol. 26, pp. 3-16

13. McDonogh, R., G. Schaule and H.-C. Flemming (1994): The permeability of biofouling
    layers on membranes. J. Membr. Sci. 87, 199-217

14. Sematech (1992): Sematech provisional test method for determining the surface associated biofilms of UPW distribution systems. Technology Transfer Number 92010958B-STD

15. Flemming, H.-C., Tamachkiarowa, A., Klahre, J. and Schmitt, J. (1998): Monitoring of
    fouling and biofouling in technical systems. Wat. Sci. Technol. 38, 291-298

16. Wasel-Nielen, J. and Nix, N (1990): Kesselspeisewasser - Erzeugung aus Flußwasser durch Ionenaustausch und Umkehrosmose. Erste Betriebserfahrungen. Vom Wasser 75, 127-141

17. Schaule, G., H.-C. Flemming and H.F. Ridgway (1993): The use of CTC (5-cyano-2,3-
    ditolyl tetrazolium chloride) in the quantification of respiratory active bacteria in bio ilms. Appl. Environ. Microb. 59, 3850-3857

18. Lowry, O.J., Rosebronkh, N.J., Farr, A.L. and Randall, R.J. (1951): Protein measurement with folin phenol reagent. J. Biol. Chem. 193, 265-275

19. Dubois, M., et al. (1956): Colorimetric method for determination of sugars and related substances. Anal. Chem. 28, 350-356

20. Blumenkrantz, N. and Asboe-Hansen, G. (1973): New method for quantitative determination of uronic acids. Analyt. Biochem. _54_, 484-489

21. Frølund, B., Griebe, T. and Nielsen, P.H. (1995): Enzymatic activity in the activated sludge floc matrix. Appl. Microbiol. Biotechnol. 43, 755-761

22. Nichols P., Henson M., Guckert J., Nivens J., and White D.C. (1985): Fourier transform-infrared spectroscopic methods for microbial ecology: analysis of bacteria, bacteria- polymer mixtures and biofilms. J. Micobial Methods 4, 79-94

23. Schmitt, J. and Flemming, H.-C. (1998): FTIR-spectroscopy in microbial and material analysis. Int. Biodet. Biodegr. 41, 1-11

24. LeChevallier, M.W., Cawthon, C.D. and R.G. Lee (1988): Inactivation of biofilm bac teria. Appl. Environ. Microbiol. 54, 2492-2499

25. Ridgway, H.F. (1988): Microbial adhesion and biofouling of reverse osmosis membra nes. In: Parekh, B.S. (ed.): Reverse osmosis technology. Marcel Dekker, New York, Basel; 429-481

26. Bott, T.R. (1990): Bio-fouling. In: Bohnet, M. (ed.): Fouling of heat exchanger surfa ces. Conf. Proc., VDI Ges. P.O.Box 1139, 4000 Düsseldorf 1; 5.1-5.20

27. LeChevallier, M.W. (1991): Biocides and the current status of biofouling control in water systems. In: H.-C. Flemming and G.G. Geesey (eds.): Biofouling and Bio corrosion in Industrial Water Systems, 113-132; Springer, Heidelberg

28. Flemming, H.-C., Schaule, G., Griebe, T., Schmitt, J. and Tamachkiarowa, A. (1997): Biofouling - the Achilles heel of membrane processes. Desalination 113, 215-225

OPTIMISING MEMBRANE PERFORMANCE
- PRACTICAL EXPERIENCES

L.Y. Dudley, F. del Vigo Pisano, M. Fazel

PermaCare International,
Aquazur Limited,
Windsor,
Berkshire SL4 3HD,
UK

1 INTRODUCTION

Europe is an expanding market for reverse osmosis (RO) technology used extensively for treating brackish water, seawater and wastewater sources. Membranes are now seen as a viable option for future potable supply in some areas of the UK.

Regulators within the European Community are imposing stricter water quality standards. To comply with these tougher regulations and discharge constraints, the use of membranes will significantly increase in this new Millennium.

The financial implications of operating reverse osmosis membrane systems below optimum performance levels can be considerable. Many large municipal plant have been constructed on a 'build own and operate' (BOO) basis and the water costs to the end user, normally a municipality or large industrial user is crucial. This cost is usually the primary decisive factor for selecting the water production process. In recent years membrane desalination has become a more viable option in many countries. Recently finished costs for BOO potable water from seawater have ranged from as low as $0.50 to $1.00 per cubic metre. Table 1 gives an example of the contribution to total operating costs for a municipal seawater system.

Table 1: *Example of Seawater RO System Costs*

	Percentage of Total Cost
Capital Recovery	18.6%
Scale Inhibition	5.8%
Membrane Replacement	5.8%
Cleaning Chemicals	1.1%
Low Pressure Pumps	4.8%
High Pressure Pumps	12.1%
Operating Labour	46.2%
Maintenance	5.6%

One the most significant contributors to the supply cost is expenditure on pumping, due to the requirement for high feedwater pressures. This however, can be minimised by the use of energy recovery systems and suitable pre-treatment to maximise the water production and minimise membrane fouling. Poor performance results in higher feedwater consumption and the need for frequent cleaning. This means increased energy and consumable costs; therefore pressure is always on the OEM to design systems to operate at optimum efficiency. There are computer packages available that can evaluate and compare the major factors contributing to total RO operating cost. These programs compare chemical dosing and energy costs and assess their ability to increase overall plant efficiency.

Figure 1: *Process Optimisation Program*

PC - Optimize

Copyright 1998, 1999 Perlorica Inc, portions PermaCare USA - without warranty of any kind.
Run Date: 28-Jun-99 [Optimization 100% Complete]

Run Identification	An Interesting Test Water		V. 1.11
	With PermaCare Program		Actual Conditions
Summary of Costs of Design Operation	$/1,000 gal Product	$/year	$/year
Energy Cost - Hi Press Pump	$ 0.387	$ 203,533	$ 207,686
Raw Water Cost	$ 1.920	$ 1,009,152	$ 1,009,152
Brine Disposal Cost	$ 0.167	$ 87,600	$ 87,600
Membrane Replacement	$ 0.254	$ 133,333	$ 400,000
Cleaning Costs	$ 0.005	$ 2,500	$ 5,000
AntiScale Cost	$ 0.136	$ 71,507	$ 78,924
Acid / NaCl Cost	$ -	$ -	$ 9,701
TOTALS	$ 2.868	$ 1,507,625	$ 1,798,063
Potential Savings with PermaCare Treatment Program - $ / yr	$ 0.55	$ 290,438	
SELECTED DESIGN CONDITIONS	VALUES	UNITS	
Permeate/Product Flow Rate	1,000.00	gpm	1,000.00
Recovery	75.0%	Percent %	75.0%
Feed Flow rate	1,333.3	gpm	1,333.3
Estimated differential pressure loss	20.0	psi	25.0
Estimated Pressure adjusted for Pi & dP	288.8	psi	293.8
Cleanings/year	2.0	number	4.0
Membrane Lifetime -years	3.0	years	1.0
Feed pH	8.10	pH Units	
Treated Feed pH	8.10	pH Units	
Acid Type Used	No Acid	Text	H2SO4 - 96%
Acid/NaCl Dose - ppm 100% basis	-	ppm	22.1
Acid/NaCl usage (100% basis)	-	lb/day	354
Acid/NaCl usage (100% basis)	-	lb/yr	129,349
Recommended PermaCare Product	PermaTreat 191		Current Product
Limiting % with Permatreat	89.3% - LSI	% of Max SI	
Recommended/Actual Feed ppm	3.40	ppm	3.75
Concentrate Concentration (ppm Inhibitor)	13.59	ppm	15.00
Daily Usage	54.42	lb/day	60.06
Yearly Usage	19,863	lb/yr	21,923
Summary at Design Conditions			
Power Required	258	kW	263
Permeate/Product produced	525,600	Gal/year	525,600
Raw Water used	700,800,000	Gal/year	700,800,000
Brine to disposal	175,200,000	Gal/year	175,200,000

The major factors, which affect membrane performance, are:
- condition of the raw water supply
- effectiveness of pre-treatment procedures
- system operating parameters
- degree of plant maintenance and continuous monitoring
- responsiveness of plant operators to significant performance changes
- the rate and degree of fouling

Membrane fouling is the most common reason for performance problems, the effects are often: - reduced membrane productivity, poor salt rejection characteristics and increasing pressure differential across the membranes.

2 MONITORING OF PRE-TREATMENT PLANT

One of the most crucial aspects for ensuring trouble-free operation is the regular monitoring of feed water condition throughout the pre-treatment system and inspection of associated equipment and pipe-work. Analysing samples of feed water at each treatment stage enables a good assessment to be made of existing plant conditions. Careful inspection of pipe-work, dosing tanks and cartridge filters can identify the presence of biofilm slime requiring sanitisation procedures to prevent continued widespread microbiological growth. If a poor scale inhibition programme is in use or the pre-treatment has failed, visual inspection may reveal colloidal particles, iron or inorganic scale on the micron cartridge filters. These are always positioned immediately prior to membrane filtration.

Closer plant monitoring could prevent many system failures. Regular monitoring should be routine with data taken and plotted either electronically or by hand. Graphs should be prepared, dated and filed appropriately with any unusual or erratic results highlighted with explanations and comments. Important parameters are - pressure drop, output, pH, temperature, conductivity, and TOC. Computerised data logging packages incorporate alarm systems to alert the operator when parameters have exceeded fixed limits.

Table 2: *Pre-Treatment Failures*

REASON FOR FAILURE	INDICATORS	MONITORING
Poor scale inhibition	Increased delta P due to scale formation on membranes, usually at the back end	Check antiscalant dosing equipment and monitor changes in water quality, softener failure
Poor sanitisation	Biofilm on pipe-work, cartridge filters & membrane. Increased delta P	Monitor sand filters, GAC, planktonic micro counts in feedwater. Check biocide dosing levels and inspect dosing tanks for biogrowth
High iron content	Visible iron loading on cartridge filters	Look for signs of pipe-work corrosion, ferric breakthrough from media beds or failure of media filters
High Organic Content	Organic matter on cartridge filter and membrane inlet	Feed water composition, review flocculation procedures, monitor feed water colour/humics & TOC

REASON FOR FAILURE	INDICATORS	MONITORING
Colloidal Breakthrough	Colloidal particles fouling the micron filters and membranes	Measure SDI & turbidity, check condition of cartridge filters, eliminate media filter fines
Fouling by GAC fines	Carbon fines foul micron filters and membrane	Check washing procedure to remove fines
Overdosed flocculant	Severe loss of flux, cationic flocculant can irreversibly foul membrane surface.	Check dosing levels and detect excess traces prior to membranes
Presence of residual chlorine	Membrane damage, high permeate conductivity and sudden increase in flux	Use Redox meters, check bisulphite dosing levels and positioning of injection point, chlorine test kit
Sand filter breakthrough	Sand and colloidal fouling of cartridge filters and membranes	Check wash procedures to remove fines
Failed acid dosing	Rapid scale formation, fast increase in delta P	Check acid desert, feed pH & increasing delta P. Inspect cartridge filters for $CaCO_3$
Climatic/Seasonal Change	High microbiological loading, biofilm slimes, on cartridge filters	Planktonic microbiological counts, look for evidence of biofilms on filters & at the membrane inlet

3 CHEMICAL PRE-TREATMENT TO INHIBIT SCALES

Feed water quality determines allowable plant operating conditions, which will dictate the optimum system recovery and overall production rate. It is necessary to make a full and detailed water analysis to identify all major anions and cations, which contribute to scale formation and general fouling. The major scaling/fouling ions are calcium, magnesium, bicarbonate, sulphate, silica, iron and barium.

Many natural waters will deposit calcium carbonate on the membrane surface if untreated. Calcium carbonate scaling potential is determined by the Langelier Saturation Index (LSI), or the Stiff & Davis Saturation Index (S&DSI) for high ionic strength waters. The risk of other scalants, such as calcium sulphate and silica are determined by measuring their ionic concentration against their known solubility products (Ksp values). Scaling results in increased pressure drop and the need for greater feed pressure to maintain constant product water output.

Dosing of chemical antiscalant reduces the risk of scaling and allows elimination of acid dosing, while maintaining plant efficiency and optimum conversion rates.

The use of an effective antiscalant will allow plant recovery to be increased to a brine LSI of up to +2.6 compared to a limitation of LSI +1.0 when using a commodity antiscalant such as SHMP or zero in an untreated system. Computer prediction programs are available from some chemical suppliers to calculate scaling potential of a range of water sources. The software allows accurate recommendations for antiscalant addition. The effect of antiscalant on scaling potential is illustrated below for a brine of LSI +2.01.The graph for the treated brine conditions indicates that the plant recovery of 70% could be increased further if plant design limitations allow. At 70% recovery 3.4ppm Antiscalant A is required to be dosed to the feedwater.

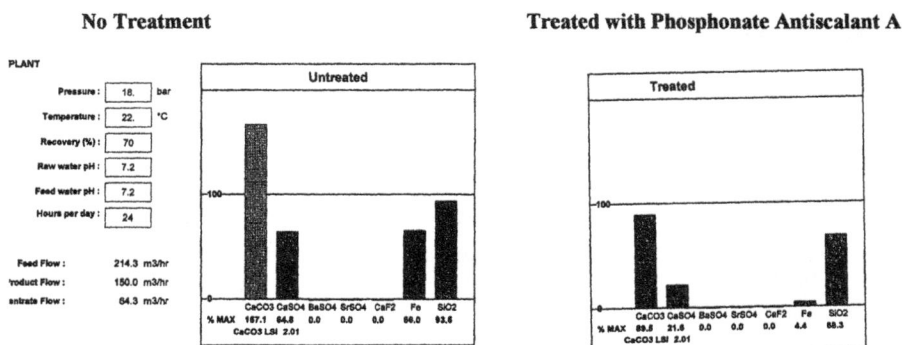

Figure 2: *Scaling Potential of RO brine with and without Antiscalant A*

In the example (Figure 2), dosing of the phosphonate based antiscalant allows safe operation of the RO plant with brine concentration of scaling species considerably in excess of normal solubility limits.

4   MINIMISING FOULING

4.1 Biofouling control

Most RO systems suffer biofouling to some degree, although this does not always severely affect performance as some biofilms remain within tolerable levels or have a high level of porosity, therefore not severely affecting the permeate flux. Biofouling potential should always be anticipated and measures taken to prevent and control biogrowth. This may require maintenance cleaning with a non-oxidising biocide or for non-potable applications intermittent biocide 'shock dosing' on-line. Laboratory analysis can be used to characterise the fouling and propose the appropriate methods of control.

Biofilm material can be scraped from fouled membrane samples for microbiological analysis consisting of basic identifications and enumeration of bacteria, fungi and yeast's. Most membrane biofilms contain both bacterial and fungal species. The physical structure of biofilms found in membrane systems can be 'gel' like or 'slimy and adhesive' with some consisting of a large ratio of polysaccharide slime to viable micro-organisms. Membrane biofilms investigated in our laboratory often contain between 10^3 and 10^8 colony forming units (cfu) of bacteria per cm^2 of fouled membrane.

Biocide Sensitivity Tests (BST's) have been used to evaluate the performance of selected biocides on sessile micro-organisms isolated from membrane foulant and determine optimum conditions for use. A quantitative suspension test is used to determine biocidal efficacy against bacteria and fungal species. Samples of the sessile organisms are obtained by swabbing the foulant from the membrane surface and spot or pour plate counts used to determine the efficacy of each biocide. The biocidal performance is expressed as percent kill for a known concentration and contact time. Non-oxidising biocide formulations are preferable due to the limited tolerance of polyamide to oxidising products such are chlorine or peracetic acid.

The following table outlines the multi-purpose use of a non-oxidising Biocide A as a periodically shock dosed biocide, sanitising cleaning agent or for membrane preservation.

Table 3: *Membrane Biocide Treatments*

Function	Conditions of Use
Intermittent 'shock dosing' on-line	Dose 60 - 80 ppm to feed water for 4-6hrs/day *Non potable applications only*
Sanitising Cleaning Agent	Recirculate 0.3% solution for 8- 10 hrs Precede and follow with alkaline surfactant
Membrane Preservative/ Biostat	Preservation period: up to 7 days: 200ppm up to 6 months: 500ppm

4.2 Maintenance cleaning

Maintenance cleaning will ensure optimum membrane lifetime and permeate production. Routine cleaning of membranes should always be carried out at a lower transmembrane pressure (TMP) than that used for water production. It is recommended that an operating pressure of less than 4 bar with minimal permeate flow is maintained for cleaning operations.

Cleaning practices should include periodic soaking of the membrane and the use of warm cleaning solutions up to 30°C. Membrane manufacturers' guidelines should always be followed regarding product compatibility and pH limits.

Cleaning of RO systems typically takes between 4 - 12 hours to perform, depending on the severity of fouling and plant size. Cleaning durations of up to 24 hours incorporating overnight soaking may be necessary if heavy biofouling is suspected. Frequency of cleaning may range from monthly cleaning cycles to an annual maintenance clean. There are many alkaline surfactants, acidic formulations and sanitising agents available in the marketplace. The complexity of the clean and number of products required for optimum cleaning conditions is wholly dependent on the composition and quantity of foulant.

5 CASE STUDY

5.1 European Paper Mill Site

Application: Production of boiler feed make-up and process water
Details: The RO plant treats town mains supply to produce 500m³/d product for industrial use. Polyamide 8" brackish water membranes are installed. Prior to our investigations, pre-treatment included - sand filtration, acidification (to pH 5.5), 5 micron cartridge filters, polymer Antiscalant X and dechlorination with sodium bisulphite. The system operates at 75% recovery.

Problem: The plant was suffering from fouling and the membranes required cleaning every 2 weeks to maintain the required treated water volume.

Fouling Investigations: Membrane autopsy revealed an orange/brown foulant covering the membrane leaves and plastic spacer. Chemical composition of the major foulants was determined as 60% organics, 11.4% calcium carbonate and 15% iron oxide.

Microbiology Results: Microbiological enumerations and identifications were performed:

	Bacterial Counts	*Fungal Counts*
	(cfu/cm²)	
Membrane	3.0×10^7	20
Plastic Spacer	8.8×10^6	2
Product Water Carrier	4.5×10^5	11

The following were identified as predominating in the foulant:
        Bacteria: rod shaped bacteria, *Arthrobacter*
        Fungi: *Trichoderma*

Biocide tests evaluated a fast acting non-oxidising Biocide B at 200ppm and 400ppm concentration with a 30 minute contact time. A 100% bacterial and fungal kill rate was achieved at the higher concentration.

Cleaning Tests: Crossflow cleaning tests using fouled membrane samples demonstrated that an alkaline clean followed by an acidic clean would successfully remove the organics, biofilm iron and inorganic scale.

Water Analysis and Antiscalant Proposal: The feedwater supply was of good quality containing negligible quantities of iron. Iron was detected in the feedwater to the RO and the planktonic counts were 1.6×10^6 cfu/ml. Inspection of the pre-treatment plant revealed corrosion of some pipework and the inside of the sand filter vessel. A computerised scaling prediction programme calculated the brine LSI using the non-acidified feed water at 20°C, pH 7.4 and 75% recovery as +1.79.

Conclusions:
- Corrosion was due to prolonged acid dosing and poor selection of materials.
- The antiscalant in use was not inhibiting scale.
- There was insufficient microbiological control at the site.

Recommendations:
- It was proposed that the polymer Antiscalant X was replaced by a phosphonate Antiscalant A. This product was to be dosed at 2.78 ppm without acid adjustment.
- It was recommended that acid dosing should be ceased to eliminate the risk of further corrosion.

- The following cleaning programme was proposed:

Step 1:	Alkaline Surfactant A	removes organics and conditions biofilm
Step 2:	Biocide B	sanitises membrane
Step 3:	Alkaline Surfactant A	removes biofilm and other organic debris
Step 4:	Weak Acid Cleaner D	removes iron oxide and inorganic scale

Outcome: All of our recommendations were followed by the site. The plant is now operating well with no indication of severe biofouling, scale or corrosion on the cartridge filters or membranes. Cleaning frequency has been reduced to every 4 months.

6 CONCLUSIONS

1. The selection of appropriate proprietary chemicals and their use in conjunction with good pre-treatment design will ensure cost-effective operation and optimise product water quality and membrane lifetime.

2. This paper demonstrates how better monitoring and laboratory investigations can identify and anticipate the cause of poor performance and allow recommendations to be made to optimise future operation. This can result in significant cost savings to the operator and end-user.

FOULING CHARACTERISTICS OF MEMBRANE FILTRATION IN MEMBRANE BIOREACTORS

M H Thomas[1], S J Judd[2] and J Murrer[3]

[1]Earth Tech Engineering Ltd,
Barnsley,
South Yorks
S75 3DL

[2]School of Water Sciences,
Cranfield University,
Beds
MK43 0AL

[3]Technology Group,
Anglian Water,
Peterborough,
PE3 6WT

1 INTRODUCTION

The combination of membrane filtration with a biological reactor is known as a membrane bioreactor (MBR), in which the membrane normally replaces the sedimentation stage of a conventional biological process. An MBR may have the membrane module submerged within the reactor (integrated) or as a separate unit (sidestream), and key aspects of these have been reviewed elsewhere [1,2], but can be summarised as follows:

- *Small footprint.* Membrane modules required to perform the separation occupy a smaller land area than the sedimentation tanks required to treat the same flow. The reactor can operate at a higher mixed liquor suspended solids (MLSS) concentration leading to a smaller volume needed to treat the same waste, leading to a further reduction in overall footprint for the system.
- *High quality effluent.* The membranes used have a low pore size (typically $0.1\mu m$ for microfiltration), which means the effluent suspended solids (SS) content is very low. The reduction in microorganisms is much greater than for conventional techniques. In an earlier study using an ultrafiltration membrane, it was found that the effluent contained no heterotrophic micro-organisms[3].
- *Better control over biological conditions.* As the solid/liquid separation is complete, all sludge can be recycled to the reactor. This means that the sludge age is independent of hydraulic retention time, giving more control over the biological process. This can lead to reduced sludge production and greater contaminant removal[4]. The high shear environment found in some sidestream MBRs can lower the average particle size in the biomass[5]. This size reduction is thought to aid mass transfer in the biomass, offering a possible explanation for improved nutrient removal rates[6].

A sidestream MBR relies on a recirculating pump to provide pressure to drive the filtration. The membranes are operated in crossflow mode, and the ability to generate a high transmembrane pressure coupled with high Reynolds numbers yields a correspondingly higher permeate flux than that attainable from the submerged configuration. The increased energy consumption is thus partly offset by the decreased membrane area requirement compared to the submerged configuration. On the other hand, the higher fluxes mean that the sidestream configuration is more prone to fouling – the accumulation of particles and soluble species at the membrane surface causing an increase in resistance to permeate flow. The propensity of membranes to foul is dependent on the feed solution, membrane type and operating conditions. The MBR should be designed to minimise fouling, as either physical or chemical cleaning adds to the cost and complexity of the process.

The major limitations of using membrane processes for this application are economic. The capital cost of membranes is high and their useful life short compared to conventional separation techniques. The driving force for membrane separation processes is pressure; hence operating costs are also high. For membrane bioreactors to be economically viable, the membrane separation step must be designed for optimum efficiency.

1.1 Clean water flux

The flux, J, of clean water across a membrane with no materials deposited on the surface can be described by [7]:

$$J = \frac{\Delta P}{\mu . R} \qquad (1)$$

where J = permeate flux $(m^3 m^{-2} s^{-1})$
ΔP = pressure drop across the membrane (Nm^{-2})
μ = absolute viscosity of the water (Nsm^{-2})
R = hydraulic resistance of the membrane (m^{-1})

1.2 Membrane fouling

Fouling may arise from particle deposits on the membrane surface, macromolecules adsorbing onto the surface or into the bulk membrane material, or pore blocking. The increase in membrane resistance is manifested as a decline in the permeate flux, as predicted by Equation (1)[8]. The solutions being filtered in an MBR contain high levels of suspended and dissolved material. The presence of these species has a significant influence on permeate flux. Earlier studies have shown the build up of membrane foulants to occur in two stages[9, 10]:

- Initial flux decline due to concentration polarisation: the presence of dissolved substances in the solution causes an accumulation of solutes on the retentate side of the membrane producing a layer less permeable to water than the membrane on which it resides.
- Long term fouling due to solute adsorption and particle deposition (gel layer and cake formation if external, pore blocking if internal): high concentrations of solutes at the membrane surface may cause precipitation to form a *gel layer*; particles in suspension are transported to the membrane surface and form deposits which reduce the hydraulic permeability and so permeate flux.

Convective transport of particles and solutes to the membrane surface is partly balanced by diffusion back into the bulk solution, possibly induced by shear. The characteristics of the system determine the balance of forces to and from the membrane surface.

1.3   Factors affecting membrane fouling

The three principal factors affecting the rate and extent of membrane fouling comprise[9]:
a) Membrane type: the membrane material, pore size & distribution and module configuration.
b) Operating conditions: factors such as pressure, cross flow velocity and turbulence.
c) Solution characteristics: the nature of both solvent and solute, concentration and nature of the bulk fluid. Published studies into the effect of increasing solids concentration of the filtered solution have reported a decline in permeate flux[3,15,16].

2   MATERIALS AND METHODS

A pilot scale rig was purpose built to test the performance of commercially available membrane modules. The mixed liquor used in the study was drawn from an existing aerobic biological reactor. A schematic of the rig is shown in Figure 1, and an analysis of the mixed liquor at different concentrations is shown in Table 1. Table 2 details the three membrane modules used during the trials.

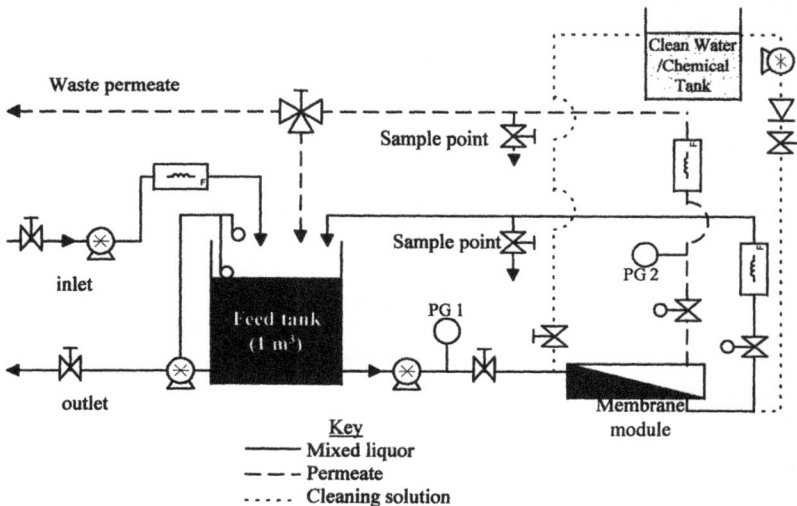

Figure 1 *Schematic of experimental rig*

Table 1   Sample analysis of mixed liquor used during trials

Measurand (mgl⁻¹)	Mixed liquor in feed tank			
SS	2500 ± 207	4680 ± 295	7230 ± 228	14900 ± 919
BOD_5	1560 ± 324	1900 ± 574	2810 ± 536	2960 ± 295
COD	2960 ± 401	4700 ± 907	7570 ± 1000	15600 ± 912
NH_3-N	6 ± 4	8 ± 4	15 ± 5	21 ± 5
TKN	12 ± 6	215 ± 106	82 ± 17	172 ± 13
DOC	7.2 ± 0.6	17 ± 14	20 ± 10	15 ± 6

Table 2   *Specifications of membranes used during trials*

	Material	Configuration	MWCO (kDa) / Pore size (m)	Total membrane area (m^2)
M1	PVDA	12mm tubular	250	2.23
M2	PVDF	12mm tubular	250	2.23
M3	PVDF	12mm tubular	0.1	2.23

3 RESULTS & DISCUSSION

3.1 Permeate water quality

The permeate water quality was consistently high. The low pore size of MF (0.05-10 m) and UF (0.001-0.1 m) membranes and the dynamic membrane formed by surface deposits ensure a high rejection of suspended solids[11]. The permeate was analysed for key contaminants during all experimental runs (Table 3). The measured values for SS and BOD_5 in the permeate were normally below the limit of detection. In previous studies using comparable membrane types analysis has shown that heterotrophic microorganisms and viruses are also rejected[3]. A reduction in the dissolved organic carbon content shows that the membranes were rejecting large organic molecules as well as particulate matter. The rejection of high molecular weight compounds and their return to the reactor are thought to explain the improved mineralisation of organic matter in MBRs[12].

3.2 Flux decline with time

Initial fouling of the membrane surface caused a severe exponential decline in flux to a semi-stable low value after a few hours. The semi-stable flux value was taken as being that after 24 hours continuous operation. Fouling also generated an increased pressure drop through the module. To enable direct comparison of the resistance due to fouling, the performance of the membranes was quantified in terms of specific flux, the permeate flux produced per unit pressure applied in $lm^{-2}h^{-1}bar^{-1}$.

Table 3   *Permeate analysis results summary*

| | Removal, % | | |
Contaminant	M1	M2	M3
SS	> 99	> 99	> 99
BOD_5	> 99	> 99	> 99
COD	> 99	> 99	> 99
NH_3-N	60 - 96	50 - 98	63 - 93
TKN	75 - 98	83 - 98	85 - 98
DOC	5 - 65	30 - 72	16 - 70

3.3   Crossflow velocity

The convection to and diffusion from the membrane surface determine the rate of fouling. The rate of convection to the membrane is a function of the permeate flux, and the diffusion away is linked to the degree of turbulence. An increase in the crossflow velocity directly increases the degree of physical scouring at the surface and hence improves back transport into the bulk solution.

The specific flux after 24 hours was recorded under the range of cross flow velocities and feed solution concentrations for each membrane listed (Table 2). The resistance of the fouling layer was found by subtracting the resistance to clean water flow from the total hydraulic resistance measured. The reduction in resistance as the velocity increases to approximately 3 ms^{-1} (Figure 2) can be attributed to increased shear induced back diffusion. The increase in resistance above 3 ms^{-1} is likely to be due to a more compact fouling layer and increased pore plugging due to the higher pressure associated with a higher feed flow rate. Published studies[10,13] have indicated that turbulence promoting baffles and other devices can reduce fouling. The use of these methods can increase back diffusion without increasing the trans-membrane pressure. This offers the advantages of increased cross flow velocity without the penalty of higher pressure and associated exacerbated fouling.

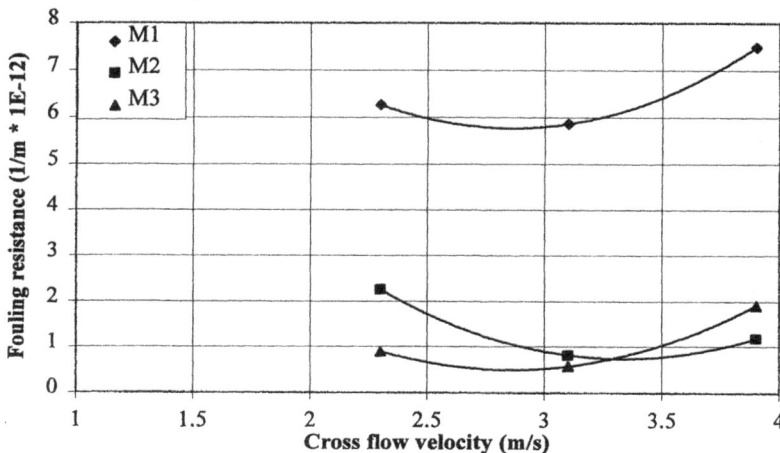

Figure 2 *Resistance of fouling deposits after 24 hours (feed concentration 2500 mgl^{-1})*

3.4   Feed solution

MBRs offer the ability to support a high biomass concentration in the reactor. However, a solution with a high concentration of suspended solids exhibits different behaviour to that of pure water. The solids increase the density and viscosity of the bulk fluid. This influences the flow regime at the membrane surface, and hence the propensity for fouling. More importantly, solids directly cause fouling through cake formation and pore blocking.

The effect of suspended solids concentration on the final specific flux (at the optimum cross flow velocity of 3.1 ms⁻¹) is illustrated in Figure 3. The data shows that fouling increases with feed solution concentration for a given operating regime. This direct comparison also illustrates the influence of membrane characteristics on fouling behaviour. The differences seen between the membranes tested in this study can be largely attributed to the surface properties of the membranes, as operating conditions were replicated as closely as possible. Published studies of membrane fouling show various relationships between stabilised flux and increasing solids concentration (Table 4).

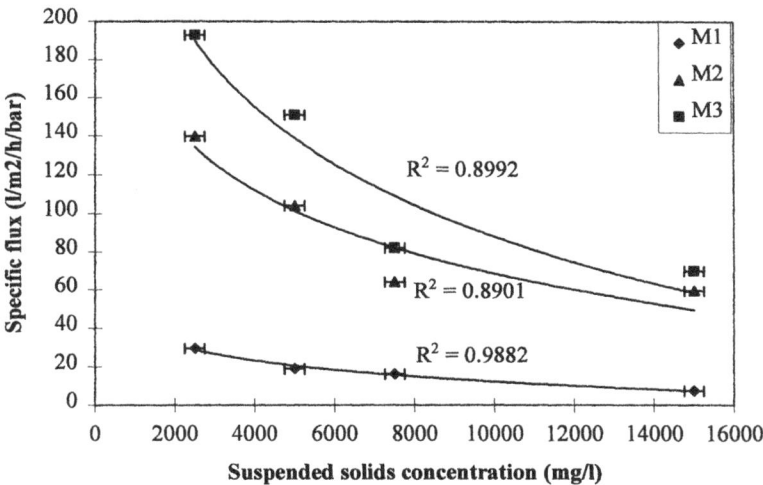

Figure 3 *Final specific flux change with feed suspended solids concentration*

3.5   Energy consumption

The energy consumption of the recirculation pumps was calculated from the measured values of flow and pressure through the system using equation (2) and (3):

$$Power \ (W \ or \ Nms^{-1}) \quad = \quad Flow \ (m^3s^{-1}) \ x \ Pressure \ (Nm^{-2}) \qquad (2)$$

The energy consumption per unit permeate is found using:

$$Energy \ consumption \ (kWhm^{-3}) \quad = \quad \frac{Power \ (W) \ / \ 1000}{Permeate \ flow \ rate \ (m^3h^{-1})} \qquad (3)$$

The results showed the minimum energy consumption per unit permeate produced to be 1.75 kWhm^{-3}, as achieved by membrane M3 at the lowest cross flow velocity employed (2.3 ms^{-1}). This illustrates the influence of the pressure drop through the system, which increases in proportion to the square of the velocity, and the higher permeability offered by the larger pore size. The energy consumption of a conventional activated sludge plant is around six times lower than the value found for membrane filtration [14].

Table 4   *Comparison of reported dependence of flux on SS*

Membrane type and pore size	Solution type and SS concentration (mgl^{-1})	Pressure (bar)	Velocity (ms^{-1})	Specific flux, J_s (lm^{-2}s^{-1}bar^{-1}) vs SS conc. (mgl^{-1})	Ref.
M1	Aerobic mixed liquor 2,500 - 15,000	0.4	3.1	J_s = -12.0 ln(SS) + 123	
M2	Aerobic mixed liquor 2,500 - 15,000	0.4	3.1	J_s = -47.4 ln(SS) + 505	
M3	Aerobic mixed liquor 2,500 - 15,000	0.4	3.1	J_s = -73.4 ln(SS) + 763	
Ceramic 4mm tubular 0.2 m	Anaerobic mixed liquor 2,500 - 22,000	0.35	2.0	J_s = -0.003(SS) + 141.6	15
Ceramic 4mm tubular 300 kDa	aerobic mixed liquor from synthetic WW 2,100 - 15,400	0.5	3.0	J_s = -54.1 ln(SS) + 621.8	3
Polymeric UF 50 kDa	Aerobic mixed liquor 5,000 - 15,000	0.3	-	J_s = -94.9 ln(SS) + 1090.9	16

4 CONCLUSIONS

1. Permeate water quality is consistently high for all membranes tested. The membrane with the highest pore size (M3) proved the most economical to operate with no detectable deleterious effect on effluent quality.
2. The resistance of the fouling layer is minimised at a cross flow velocity of approximately 3ms^{-1} for the membrane modules tested.
3. The semi-stable permeate flux after 24 hours continuous operation shows a logarithmic decline with increasing suspended solids concentration. The relationship between semi-stable flux and suspended solids concentration of the feed solution has been compared for the membranes used in this trial and other published reports. The wide differences between findings indicate that membrane surface properties play a large part in the fouling characteristics.
4. The lowest energy consumption for the filtration was 1.75 kWhm^{-3} and was achieved at the lowest cross flow velocity using the membrane with the largest pore size.

5. The high energy consumption of a sidestream MBR must be justified by a need for the particular benefits offered by the process. This need could arise from strict effluent control, high land costs, prohibitive sludge handling and disposal costs or high strength wastewater requiring specific biological conditions.

References

1. T. Sato and Y. Ishii, *Wat. Sci. Tech.*, 1991, **23**, 1601.

2. K.-H. Choo and C.-H. Lee, *Wat. Res.*, 1996, **30**, 1771.

3. N. Cicek, H. Winnen, M. T. Suidan, B. E. Wrenn, V. Urbain and J. Manem, *Wat. Res.*, 1998, **32**, 1553.

4. E. Trouve, E., V. Urbain and J. Manem, *Wat. Sci. Tech.*, 1994, **30**, 151.

5. A. D. Bailey, G. S. Hansford and P. L. Dold, *Wat. Res.*, 1994, **28**, 297.

6. American Water Works Association Research Foundation (AWWARF), Lyonnaise des Eaux (LdE), Water Research Commission (WRC) of South Africa, *Water Treatment Membrane Processes*. New York: McGraw-Hill, 1996.

7. M. H. Al-Malack and G. K. Anderson, *Wat. Res.*, 1997, **31**, 3064.

8. G. Belfort, and F. W. Altena, *Desalination* 1983, **47**, 105.

9. A. G. Fane and C. J. D. Fell, *Desalination* 1987, **62**, 117.

10. J. A. Howell and S. M. Finnigan, in *Effective Industrial membrane processes - Benefits and Opportunities*, Elsevier Science, 1991, p.49.

11. M. C. Porter, *Microfiltration. Synthetic Membranes: Science, Engineering and Applications*, D. Reidel, 1986, p225.

12. K. Brindle and T. Stephenson, *Biotechnology and Bioengineering*, 1995, **49**, 601.

13. R. W. Field, D. Wu, J. A. Howell and B. B. Gupta, *J. Membrane Sci.* 1995, **100**, 259.

14. T. Ueda, K. Hata and Y. Kikuoka, *Wat. Sci. Tech.*, 1996, **34**, 189.

15. A. Beaubien, M. Baty, F. Jeannot, E. Francoeur, and J. Manem, *J. Membrane Sci.* 1996, **109**, 173.

16. Y. Magara and M. Itoh, M., *Wat. Sci. Tech.*, 1991, **23**, 1583.

CLEANING OF MEMBRANES IN WATER AND WASTEWATER APPLICATIONS

Ralf Krack, Technical Manager Membrane Cleaning / Water Treatment

Henkel-Ecolab GmbH & Co.
OHG, P.O. Box 130406,
D-40554 Düsseldorf

1 INTRODUCTION

For more than 25 years membrane filtration processes are used as state of the art technology in many different applications e.g. seawater desalination and applications in food processing industry and biotechnology. With increasing costs for water and wastewater a rapidly increasing tendency of using membranes can be observed in Central Europe.

1.1 What is water?

Water in the definition of the terms of this topic can be sea water with high amounts of sodium chloride, surface water with varying amounts of minerals and organic ingredients like humic acids but as well normal potable water already treated water from the water works.

1.2 What is wastewater?

Wastewater can be the effluent water from an industrial company discharged to the sewer, this is known as "end of pipe" treatment. Here, a wide range of compounds are present in varying concentrations. Wastewater can also be water treated using biological processes. Here, bacteria, residues of flocculants and sewage sludge can present. Wastewater can also be a part of a production plant e.g. a washing machine or bottle washer. In this case an influence on the composition of the wastewater can be made and often not only the water can be reused but useful products, which can have an added value for use in the production process again.

1.3 Which membrane processes are in use?

Depending on the feed and quality demands of the filtrate any filtration process ranging from microfiltration up to reverse osmosis involving a wide variety of membrane materials can be found. If a laundry or a car producer wants to meet the consent limits for the discharge of wastewater a simple microfiltration or ultrafiltration unit may be appropriate. If they want to re-use the water a nanofiltration or a reverse osmosis plant is needed. The composition of the wastewater in terms of viscosity, turbidity, fibres and abrasive materials determines the type of membrane e.g. tubular, hollow fibre, flat sheet or spiral wound. The permeate quality, stability of the membrane, lifetime and amount of investment could determine if a ceramic or polymer membrane is chosen.

2 CLEANING

2.1 Why is cleaning of membranes essential?

Membrane surfaces - for potable water production - have to be cleaned like any other surface that are in contact with food or pharmaceuticals to guarantee a high quality product as well as hygienic standards. In addition a good cleaning procedure is also essential for the functionality and capacity of the membrane plant, even if bacteriological requirements are as low as in most wastewater applications.

2.2 What are the requirements for a sufficient cleaning?

A sufficient cleaning can only be guaranteed, if the applied cleaner is optimised for removing the soil and does not adversely influence the membrane characteristics. As a simple example, the use of a caustic soda applied in highest concentration will not be able to remove a mineral scaling.

Additionally the physical aspects of membrane cleaning like temperature, mechanical forces, time and last but not least the chemical activity are points of major concern in membrane cleaning processes. Even the most important "solvent" in membrane cleaning - the water - should be of the best quality; otherwise the cleaning itself could become a cause for fouling.

2.3 Physical aspects of membrane cleaning

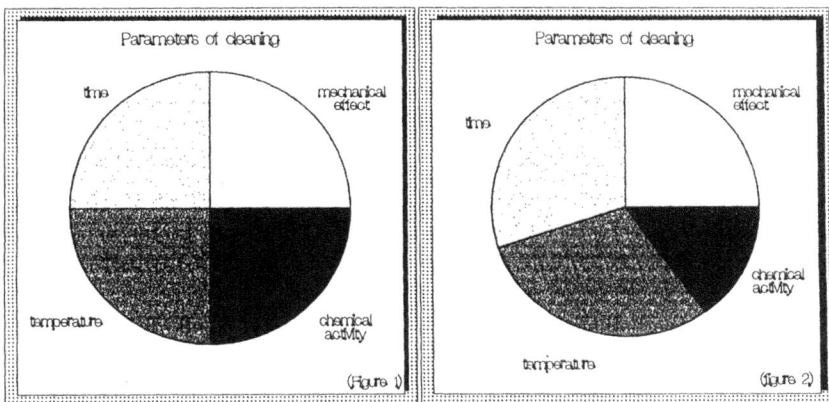

The "Sinnersche Kreis", Figure 1 and 2, is one of the most important and well known descriptions for cleaning in general that can be used for membrane cleaning. To get the optimum cleaning result, time, temperature, mechanical force and the right chemical activity is needed. If one of the parameters is decreased at least one of the others has to be increased in many cases two or all three have to be increased.

2.3.1 Temperature It is well known that an increase of the temperature increases the reactivity of processes. This is the same for cleaning processes. As a major value it can be stated that an increase of 10°C doubles the cleaning activity. Nevertheless depending on the different foulants a minimum temperature might be necessary. Many polymer membranes unfortunately allow only a certain temperature. Any exceeding of these limits can cause sever damage of the membrane. For the user it is complicated to understand the reasons for different limitations even if the membrane material applied is the same (Figure3).

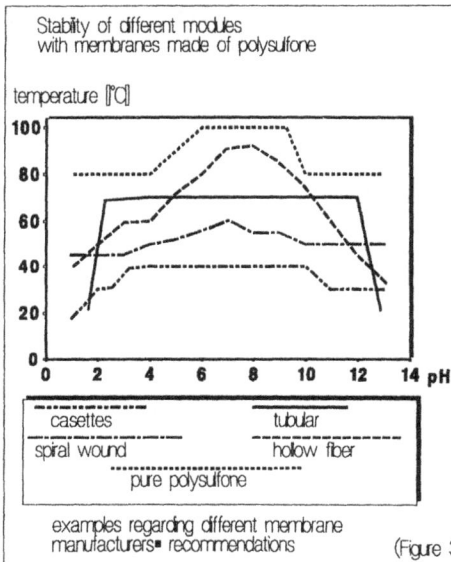

Stability of different modules with membranes made of polysulfone

temperature [°C]

casettes — tubular
spiral wound — hollow fiber
pure polysulfone

examples regarding different membrane manufacturers' recommendations　　　(Figure 3)

Even if the membrane materials are the same, the spacer material, the support material, and the glue may influence the module stability. For that reason it is important to have all data available so as to make good cleaning recommendation

Especially in case of water and wastewater treatment plants there is no option to heat up the cleaning solution. That is one of the major differences compared to food application where on very large plants with a cleaning volume of 5000 litres a possibility for heating is state of the art. If there is a wastewater treatment plant producing a latex emulsion at 40°C the pre rinse should already be done at the same temperature. If the pre rinse can be done only with cold water the emulsion may break down and will lead to much more difficult cleaning conditions.

In some cases cleaning might be impossible. Comparable may happen if a wastewater with high amounts of proteins is treated. High concentrated proteins can be liquid at high temperature but can become solid if they are cooled down.

2.3.2 Mechanics The relationship between flow and cleaning results are well known from different models and practical experiences. Higher flow rates lead to higher turbulence and to better emulsifying and dispersing properties. The increase of flow rates is limited by pump capacity, mechanical stability and of course in the economics. During membrane cleaning it is important to watch the pressures both parallel to and vertical to the membrane surface. Therefore it is necessary to illustrate the different types of fouling depending on the different membranes. On ultra- and nanofiltration membranes as well as on reverse osmosis membranes mostly the fouling or scaling layer is directly on the surfaces (figure 4). Using microfiltration membranes as well as surface fouling an inner porous fouling appear (figure5).

During cleaning of UF, NF, and RO plants the filtering capacity should be as low as possible, otherwise it is possible that a secondary membrane builds up during cleaning. For cleaning MF plants it is the best to clean first with a low pressure to take away the loose cake on the surface then discharge this first dirty solution and finally change to a higher filtration capacity to clean the pores of the membranes.

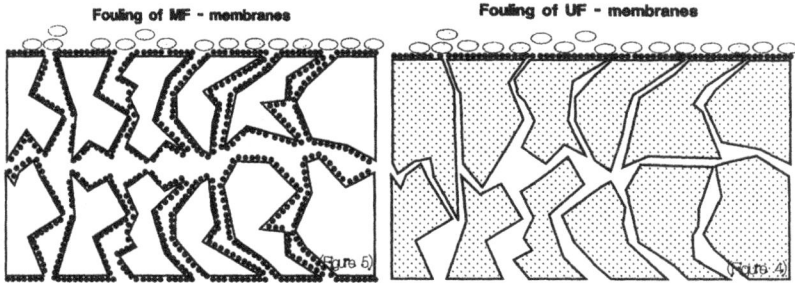

An often discussed cleaning procedure is backflushing (many membranes and modules cannot be back flushed). In practice we have learned that back flushing is very effective in improving membrane capacity when a hard cake layer is present, although a totally clean situation will never be achieved. By using backflushing, as a method of removing crude surface fouling, care must also be taken that the monomolecular and oligomer fouling is removed. The removal of monomolecular and oligomer fouling is independent of the direction of flow; chemical aspects are most responsible for the clean. Another method is a cleaning from the permeate to the concentrate side. This may be an alternative process on membranes which allow this type of cleaning but it will only function well if the dirt is discharged and not recirculated. If the cleaning solution is recirculated then emulsified fouling will come from the feed side (dirty side) to the permeate side (clean side) and may foul the membrane from the permeate side. However even if the permeate side is not fouled, an additional risk is the possibility of contamination with bacteria.

2.3.3 Time Generally it is stated that the effect of an increased time increases the cleaning efficiency. That is valid for many cases. Especially in membrane cleaning the practical experiences have shown also different effects. During cleaning our membrane system is still able to filter out some only emulsified or dispersed soils.

The exact time for a cleaning can only be given by the practical experience or by direct control of the cleaning process.

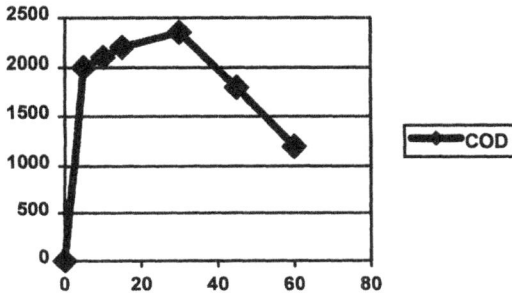

Figure 6. *Effect of cleaning time*

2.4 Chemical aspects of membrane cleaning

Generally the chemicals applied for membrane cleaning are as important as the water quality. Many users are not aware of the problems that can appear if the wrong chemicals or water quality is applied.

2.4.1 Water quality   Water itself is a very good solvent and also a good cleaner. Water used in membrane cleaning should be of good quality. Any impurities can be filtered out during cleaning and may block the membrane rather than cleaning them. Prefilters should remove any suspended solids. The bacteriological contamination should be at a very low level to prevent biofouling. Several metals form salts, which precipitate especially iron and manganese, these; together with silicates can form insoluble salts. The normal deposition of non-silicate salts of iron and manganese may be removed but silicates can only be removed by hydrofluoric acid. This is not a viable proposition as, using this, most membranes and nearly every plant would be destroyed, not including any human risks. Water hardness is not a problem. With formulated detergents, both acidic or alkaline hardness can be complexed or removed. Very often demineralised water is used, being produced by ion exchange, RO plants or evaporators. Water from evaporators can sometimes cause problems on membranes, as antifoamers are often used in defoaming evaporated solution. These antifoamers can escape into the vapours. Contamination with even very small amounts can decrease the membrane capacity with time.

Measuring the silt density index (SDI) can easily test the water quality. There are published values for water quality from which it can be determined whether or not they can be treated by RO plants. Here the SDI should not be higher than 5. For cleaning of membrane plants the SDI should be less than 3, otherwise problems may occur.

2.4.2 Influence of the soil   The soil itself has a direct influence on the chemicals, which should be used for cleaning the membranes. For membrane cleaning it is necessary to determine the type of soil present much more than for usual stainless steel surfaces.

2.4.2.1 Fat, oil and other hydrophobic fouling   The removal of hydrophobic fouling like fat from hydrophobic surfaces like organic polymers is more difficult than from steal or glass. This is depending on the hydrophobic characteristic of fat that adsorbs more strongly on the membrane surface. The removal can be done with surfactant based products above the melting point of the fat or grease in aqueous solution. Below the melting point it is nearly impossible to remove fat and grease residues without using a real solvent like ethanol. The surfactants must be chosen precisely, because they have to be compatible with the membrane, the spacer and the support. Surfactants and defoamers used in normal CIP detergents or washing powders are often not compatible with membranes even if temperature and pH limitations are suitable. Some membrane manufacturers have approved normal washing powders in the past and could reach quite good cleaning results. However a normal washing powder is usually changed at least once a year and of course nobody will take care that the old version was approved on a membrane system.

2.4.2.2 Proteins   Alkaline detergents remove proteins best, this is well known from practical experience. The higher the pH the faster the protein hydrolysis and the better the solubility. Protein solubility is poor in the neutral pH range. At a pH of 4-5 milk proteins for example are denatured and precipitated. The initially used pH-sensitive membranes like cellulose acetate led to the development of enzymatic detergents. Many of today's commonly used membranes are pH resistant. Initially, only inorganic ceramic membranes were pH stable. Latterly also organic polymers like polysulfone, polypropylene and polyvinyldifluoride with high pH stability were developed. The exact specifications vary from manufacturer to manufacturer. The best cleaning process is not achieved by attaining the correct pH alone. A pH of 11-11.5 is very often the limit; even a small amount of caustic may be enough to reach this value but not sufficient to clean well. Additionally to the alkalinity there is a need for dispersants, emulsifiers, soil carrying agents, stabilisers for hardness salts, buffering systems and available chlorine or oxygen as cleaning boosters. Enzymatic cleaners are usually applied if the pH limitation is at or below 10 or if a high level of dirt is present.

   2.4.2.3 Minerals and salts   Most of the mineral scale can be easily removed by an acid cleaning with organic or inorganic acids. Alkaline products containing complexing agents can remove some others. In cases of silicate and sulphide residues it is best to prevent than to clean.

   2.4.2.4 High molecular polysaccharide, EPS extra cellular polysaccharide Polysaccharides, in particular, cause severe problems if they appear on non-oxidising stable membranes. Here two typical phenomena appear. The one is that EPS is directly fixed on the membrane surface and lead to a strong decrease of the capacity. The other especially in spiral wound membrane systems appearing is that the EPS settle down in the spacer material and rise the difference pressure. Both can lead to a loss in the performance of the membrane system. The cleaning process in such cases has to be adapted individually. One successful process to decrease the difference pressure is the enzymatic cleaning followed by an acidic sanitation and another enzymatic cleaning step (Figure 7).

Development of difference pressure with two different cleaning processes

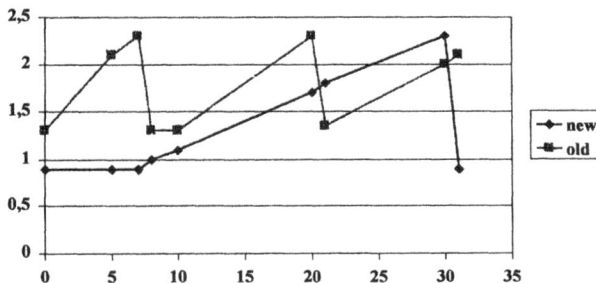

(Figure 7)

Whereas with the old cleaning procedure a cleaning became necessary every 7 to 10 days, the new enzymatic based process could lengthen the cleaning cycles to around 30 days. Trials that were done during optimisation of the plant performance with additional disinfection steps could kill most of the biofilm producing bacteria but did not remove them from the membrane and spacer material. So the dead bacteria became ideal food for new bacteria and speeded up finally the developments of new colonies.

   2.4.2.5 Cleaning agents from industrial companies   Many wastewaters contain cleaning agents from the factories that may influence the behaviour of the membrane system. If only part of the total wastewater of a factory is treated then it is easier to control what detergents are used. Within screening test it can be controlled what influence the detergent has on a membrane system. If there is a decrease in capacity another detergent for the factory cleaning has to be applied. The producer of the cleaning products should be able to offer a product, which fulfil as well the cleaning in the factory as the treatment by a membrane unit.

In some cases the pH during recycling is of major interest. There might be a cleaning agent that can be recycled at a pH of 8 without any problems but if decrease the pH to value of 5 a blocking of the membrane may appear. It is well known that the temperature has a strong influence on the capacity of membrane systems. Usually by increasing the temperature the capacity will increase as well. If a wastewater with some cloud point depending antifoamer is treated the reaction may be the opposite. Meanwhile the product is clear and good soluble at 25°C it becomes insoluble at 45°C and decrease the capacity of the membrane system or even block it.

3 CONCLUSION

Because of the costs and inconvenience incurred when changing damaged or blocked membranes in large filtration plants, much care should be taken. In any case it is recommended to do pilot studies before using a large membrane plant. Such studies should include the membrane cleaning aspects as well as any changes that might be necessary to give excellent results. Today's membrane technology is in most of all cases able to produce clean and healthy water independent of the feed quality. To make the processes beneficial to the user, the membrane manufacturer, the plant supplier and the producer of detergents these all have to work together from the very beginning of new projects.

Water Reuse

WATER REUSE FOR THE NEXT MILLENNIUM - MEMBRANE TREATMENT AT THE MILLENNIUM DOME

J. H. Khow[1], A. J. Smith[1], A. Rachwal[1], A. Donn[2] and C.V. Meadowcroft[2]

[1]Thames Water R & D
Spencer House,
Manor Farm Road
Reading RG2 0JN

[2]Leopold - PCI Membranes
Laverstoke Mill,
Whitchurch
Hants RG28 7NR

1 INTRODUCTION

As we move into the new millennium, increasing population and water use will lead to increased demands on water resources world wide. South East England is no exception, with water demand increasing by approximately 2% per year. The long term trend of climate change towards hotter summer temperatures and drier winters will challenge conventional surface and groundwater resources which require the winter rains to top up reservoirs and groundwater levels. Sustainability of water resources is a key factor in maintaining supplies to our customers.

Thames Water's water resources strategy is based on the management and enhancement of conventional resources, the investigation and use of alternative water sources where appropriate, and water conservation through the application of new technology and the understanding of customer demands and behaviour.

In accordance with this strategy, Thames Water, in association with the New Millennium Experience Company (NMEC), have implemented the first major in-building recycling scheme in the UK. The wide ranging objectives of the scheme are to promote sustainable water use, demonstrate and research water recycling technologies, evaluate water efficient appliances and investigate public attitudes to water recycling initiatives.

The water recycling system has been constructed at the Millennium Dome, Greenwich which is to be the focus of the country's millennium celebrations. At a cost of £758 million, the Dome is the largest building of its kind in the world with a perimeter of 1 km and height of 50 m at its centre. An estimated 12 million people will visit the Dome during the year 2000[1].

As part of the Dome's environmental and water management strategy Thames Water's Water Reclamation Plant will supply 500 m^3 per day of reclaimed water to flush all of the WC's and urinals on site. Water is reclaimed from three sources, greywater from the hand washbasins in the toilet blocks, rainwater from the roof of the dome and groundwater from the chalk aquifer below the site.

2 WATER SOURCES AVAILABLE FOR RECYCLING

2.1 Grey Water

The Dome has six core buildings that house the toilet blocks. Greywater will be collected from the hand basins in these blocks. The expected 35,000 - 55,000 visitors each day and staff are predicted to use, on average, 120 m³/d of hand basin water which will then be treated and reused. This figure is the best estimate based on available information, including the Chartered Institution of Building Service Engineers (CIBSE) design recommendations[2] and studies of hand basin water demand in domestic situations[3,4].

2.2 Rainwater

The 100,000 m² surface area of the Millennium Dome roof provides the opportunity to capture large quantities of rainwater run-off during rainfall events. Rain runs from the roof through specially designed hoppers that feed into the surface water drainage system. Although the potential volumes are high, the constraints of the site mean that the area available for storage is restricted. A maximum flow of 100 m³/d is pumped from the main surface water collection system into the surface water treatment system.

2.3 Groundwater

London, like many other Metropolitan cities, has a problem with rising groundwater. Historical over-pumping for industrial and commercial supply from the chalk had reduced natural groundwater levels by over 50 m. However, since 1970's decline in these pumping rates due to the changing industrial base has allowed a rapid recovery of water levels. Groundwater levels have already risen by over 35 m over the last 20 years with rates of rise under the City of London and Westminster of over 2 m per year.

GARDIT (General Aquifer Research and Development Investigation Team), a multi-organisational grouping set up to look at the problem of London's rising groundwater has identified the Greenwich peninsular as a potential borehole location where pumping from the aquifer would be beneficial. A 110 m deep borehole has been drilled on the Millennium Dome site to provide the 600 m³/d required and also to contribute to the de-watering of the aquifer.

3 PROCESS SELECTION AND PILOT PLANT STUDIES

Due to the tight deadlines the project required fast tracking specific pilot trials for process selection. The treatment processes were chosen with additional regards to security of supply and future flexibility.

3.1 Grey Water Treatment

The greywater feed quality was established from previous work[5,6] and trials carried out using the handwash basins in the Thames Water R&D office complex. A survey of soaps and detergents' manufacturers confirmed that all handwash soaps are significantly biodegradable and that the majority of modern soaps supplied for large scale public applications include synthetic surfactants[7]. The synthetic feed quality used for pilot trials is shown in Table 1.

Table 1 *Synthetic greywater feed quality*

Determinants	units	Synthetic Greywater
Total BOD	mg/l	59
Soluble BOD	mg/l	37
Total COD	mg/l	153
Soluble COD	mg/l	109
TOC	mg/l	34
Suspended solids	mg/l	18
Anionic surfactants	mg/l	6
Non-ionic surfactants	mg/l	2.1
NH_4 - N	mg/l	<0.05*
Total Kjeldahl N	mg/l	3.8
Total P	mg/l	<0.2*
Hardness as $CaCO_3$	mg/l	273

(* - limit of detection)

A literature survey showed that world-wide recommendations for total BOD for flushing purposes ranged from 20 mg/l to 2 mg/l[8-10]. From Table 1, it can be seen that greywater requires further treatment prior to reuse, even for toilet flushing. The relatively high BOD concentration will lead to biological growth in the reclaimed water distribution system, with the added risk of odour formation. The nature of the source of the greywater also gives a high risk of contamination with pathogenic micro-organisms[4].

Considering previous work, the proposed treatment for the hand basin greywater is biological treatment using a Biological Aerated Filter (BAF) followed by membranes[11]. In order to ascertain the optimum configuration of these processes a range of pilot trials were undertaken with specific emphasis on BOD removal[7]. The performance of a pilot scale BAF, followed by a variety of membranes were investigated using the synthetic greywater. The BAF comprised of two downflow columns, each with a diameter of 150 mm, a height of 2 m, and a total bed volume of 0.036 m³. Lytag pulverised fuel ash media was used in both columns. The membrane test rig used for treating the BAF effluent consisted of six single tubes, with a total membrane area of 0.22 m². It was operated in a batch mode with a variety of PCI's ultrafiltration (UF), nanofiltration (NF) and reverse osmosis (RO) membranes.

The results in Table 2 showed that BAF is capable of achieving a total BOD of between 20-25 mg/l in the effluent from an initial value of 60 mg/l, yielding a reduction of 58-67% total BOD. All of the membranes tested achieved permeates containing between 2.3 to 10.6 mg/l total BOD, within the 2-20 mg/l recommended range. In order to minimise biological growth in the distribution system, tight UF membranes, type 2 and 3 were probably sufficient for hand basin greywater treatment. However, before a final decision on membrane type was made, the quality of the other source waters had to be considered.

Table 2 *BOD reduction following BAF and membrane treatment*

	greywater influent	BAF effluent	UF open → 1	UF tight → 2	NF → 3	RO → 4	5	6
Total BOD (mg/l)	60	20-25	10.6	5.1	6.3	3	2.3	4.8
Soluble BOD (mg/l)	30	6-10	8.3	4.7	5.8	3.1	2.4	4.7

Key:

membrane type	material	rejection characteristics	product name
1	polyvinylidenoflouride	MWCO 200,000 Daltons	FP 200
2	modified polyethersulphone	MWCO 6,000 Daltons	EM 006
3	polyethersulphone	MWCO 4,000	ES 404
4	polyamide film	75% CaCl2 rejection	AFC 30
5	polyamide film	80% NaCl rejection	AFC 80
6	cellulose acetate	90% NaCl rejection	CDA 16

3.2 Rainwater Treatment

As the Dome has not been constructed at the design phase the roof runoff quality was not available. However as an indication of water quality to expect, the composition of roof runoff from buildings at a major European city has been reported[12] and is shown in Table 3. The rainwater analysis highlighted a significant amount of hydrocarbon and heavy metals such as cadmium, copper including lead, probably attributed to vehicle exhaust pollution. The wide variation between the minimum and maximum concentration of the measured parameters is probably due to the intensity, duration and period between rainfall event. Due to the potentially high concentrations of contaminants especially after a long dry spell, the precaution of avoiding the higher contaminated "first flush" was included in the design.

Constructed wetlands are commonly used for surface water treatment. Micro-organisms that exist in the roots of the reeds break down micro-pollutants efficiently while heavy metals are absorbed into the root system. The process can cope with flow surges, is economical to operate and maintain and will not need to be replaced over the duration of the project. Reed beds were also chosen for their aesthetic value and to demonstrate the performance of a 'low-tech' natural treatment.

Table 3. *Composition of Roof Runoff in Paris*

Parameter	Units	Min.	Max.	Median
Suspended Solids	mg/l	3	304	29
COD	mg/l	5	318	31
BOD5	mg/l	1	27	4
Hydrocarbon	μg/l	37	823	108
Cadmium	μg/l	0.1	32	1.3
Copper	μg/l	3	247	37
Lead	μg/l	16	2764	493

3.3 Groundwater Treatment

Analysis of the groundwater is shown in Table 4. High concentrations of sodium chloride, hardness and hydrocarbons were found in solution, with hydrogen sulphide gas also detected. The inorganic and organic water quality results suggest that the borehole water is a mixture of chalk groundwater and Thames tidal river water together with contamination from industrial pollution, possibly from the industry previously located on the site.

Hydrogen sulphide can be removed from water by aeration, chemical oxidation and adsorption onto carbon. Chemical oxidation was selected because of the advantage of converting the sulphide to sulphate which is innocuous. Hydrogen peroxide was chosen as the oxidant as it does not form any by-products. The two possible oxidation reaction reported[13] are:-

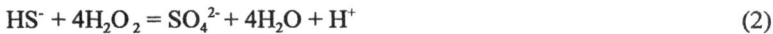

$$H_2S + H_2O_2 = 1/x \, S_x + 2H_2O \qquad \text{where frequently } x = 8 \qquad (1)$$

$$HS^- + 4H_2O_2 = SO_4^{2-} + 4H_2O + H^+ \qquad (2)$$

A stoichometric excess of hydrogen peroxide is required to effect reaction 2 within 15 minutes[13]. Following chemical treatment the groundwater is passed through granular activated carbon columns to effect removal of hydrocarbon. The remaining salinity, hardness and residual pollutants would be removed by RO.

Table 4 *Analysis of Groundwater Quality*

Parameters	Units	Minimum	Maximum	Average
pH		7.3	7.5	7.4
Conductivity	µS/cm	4090	4665	4342
Colour	Hazen	1	5	2.7
Turbidity	Formazin	1.3	20.8	8.6
TOC	mg/l	2.4	3.2	3
TDS, 180°C	mg/l	2419	2943	2705
Hardness as $CaCO_3$	mg/l	714	962	783
Total Iron	µg/l	161	223	200
Dissolved Iron	µg/l	12	12	12
Manganese	µg/l	4	127	39
Calcium	mg/l	149	159	155
Magnesium	mg/l	87	95	90
Sodium	mg/l	680	869	756
Alkalinity as $CaCO_3$	mg/l	275	324	293
Sulphate	mg/l	165	249	217
Chloride	mg/l	1208	1470	1309
Nitrate as N	mg/l	1	1	1
Fluoride	µg/l	734	772	755
Toluene	µg/l	7.2	8	7.6
Xylene	µg/l	14.7	16.4	15.6
Benzene	µg/l	20	21	20.5
Ethylbenzene	µg/l	8	8.4	8.2
Sulphide	µg/l	1.1	3	1.6

4 TREATMENT PROCESS FOR THE DOME

4.1 Description of Overall Treatment Process

The final treatment processes selected is shown schematically in Figure 1. Grey water is collected in a balancing tank prior to biological treatment through a BAF at a hydraulic loading rate of approximately 2 $m^3/m^2/d$ to reduce the BOD load.

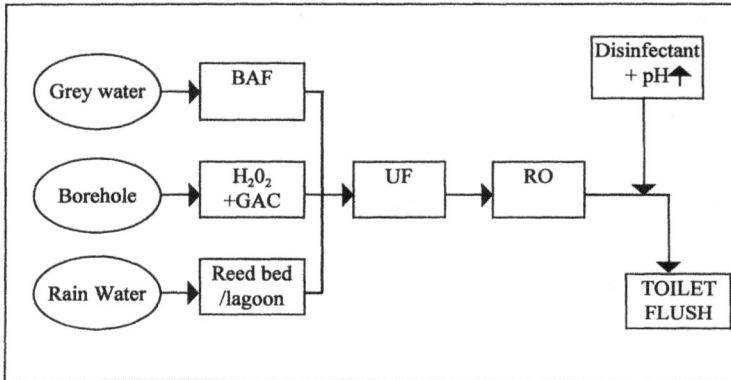

Figure 1 Process schematic of the grey water recycling treatment process

Groundwater is pumped out of the borehole, dosed with hydrogen peroxide and held for a 15 minute contact time to oxidise the hydrogen sulphide and the dissolved ferrous to ferric iron. The water then passes through two granular activated carbon (GAC) absorbers arranged in parallel to remove organic contaminants. Each absorber contained 5 m^3 of carbon and was designed to handle a hydraulic flow rate of 10 $m^3/m^2/h$.

The rainwater is passed through a series of reed beds and a lagoon which form part of the landscape design. The first reed bed is designed for storm water treatment at 5.0 $m^2/m^3/d$. Following this is a storage lagoon of approximately 400 m^2. The second reed bed will perform a tertiary treatment function. Each reed bed has an area of 250 m^2 and is designed for a maximum flow of 100 m^3/d. All flows beyond this will be discharged to the River Thames. The beds are approximately 0.6 m deep and have a 0.5% gradient. The media is washed river gravel of 5-10 mm, planted with the common reed, *Phragmites australis,* and grown from an appropriate seed to ensure salt tolerance[14].

All three influent process streams combined in a single balance tank where they are mixed and pumped to the UF plant. There is a hierachy for the treatment of the water sources. Greywater is treated in preference to rainwater and groundwater is the least preferred source.

UF was selected to remove pathogens and particulate materials so that the RO plant can operate optimally. RO is required due to the salinity and high level of hardness of the groundwater. Both the product streams from the BAF and Reed Beds are treated through the UF/RO membrane systems to remove the potential for biological growth to a greater degree and provide extra security within the reclaimed water distribution system within the Dome.

The RO product is disinfected with a residual chlorine dose, re-hardened to reduce its corrosion potential on plumbing fittings and stored before being pumped to the Dome for flushing the WCs and urinals when required. The water quality will exceed current draft UK standards for flushing[15].

4.2 Integrated Membrane Treatment

4.2.1 Ultrafiltration   The fully automatic UltraBar UF hollow fibre membrane plant was built by PCI Leopold. The schematic of the UF process is shown in Figure 2. The plant is designed to treat a feed flow of 700 m³/d and to achieve a recovery of 85%. The membranes have a nominal pore size of 0.01 micron and are made from hydrophilic polyether sulphone/polyvinyl pyrrolidone blend. The single stack unit consist of fourteen standard RO pressure vessels fitted with twenty-eight hollow fibre membrane elements each with an area of 21m².

A programmable logic controller (PLC) automatically executes a pre-selected filtration, backwash and cleaning programmes. In the process mode, feed water is pumped at typically 0.5 bar through a 100 micron backwashable screen and fed to both ends of the pressure vessel in a dead end mode. Under hydraulic pressure contaminants larger than the membrane pore size are rejected and are collected on the inside lumens of the fibres. The UF filtrate is collected in a balance tank, which also feeds the RO process. At the end of the filtration cycle a small volume of UF permeate is used to reverse flush the membranes to remove the filtered contaminants from the membrane modules. Periodically 100 mg/l of chlorine is dosed into the backwash to inhibit bacterial growth in the system. The UF plant is also designed to allow dosing with acid and caustic to control mineral scale or organic fouling.

Due to the extremely low tolerance of the RO membranes to chlorine, the UF filtrate flow to the RO plant is dosed with sodium bisulphite for a period of time after each chlorine backwash to ensure complete removal of free chlorine. A redox meter is fitted on the RO plant upstream of the membrane modules to shut down the plant if any free chlorine is detected to prevent membrane damage.

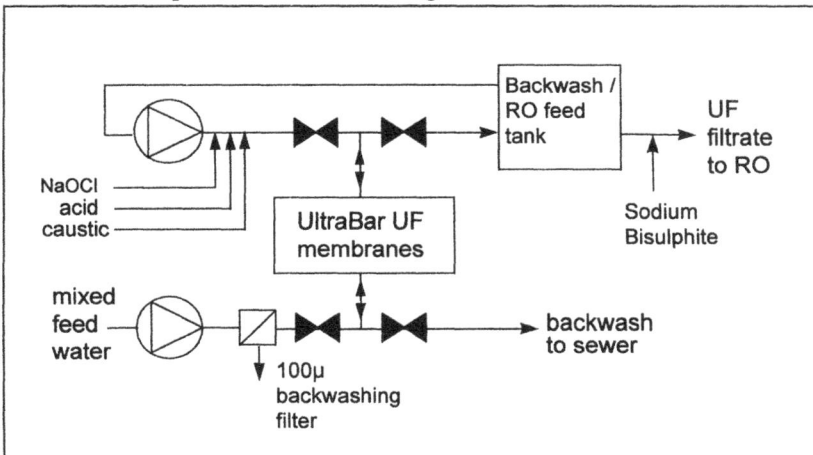

Figure 2 Schematic Diagram of the UF Process

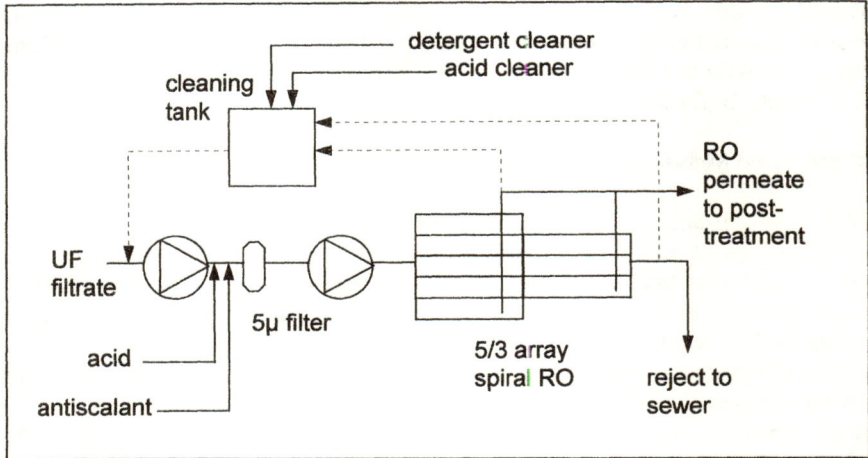

Figure 3 Schematic process for the RO plant

4.2.2 Reverse Osmosis   The fully automatic RO plant was supplied by PCI Leopold. The plant is designed to treat a feed flow of 600 m³/d and to achieve a recovery of 85%. The schematic flow diagram of the RO process is shown in Figure 3. The UF filtrate is dosed with sulphuric acid to acidify the water in order to control carbonate scaling during the concentration process. Houseman Permatreat 191 antiscalent is dosed as a scale inhibitor to prevent precipitation of insoluble salts. The feed water is then prefilter through a nominal 5 micron cartridge filter and the pressure boosted prior to filtration across the RO membranes arranged in a 5/3 array to achieve the optimum hydraulic efficiency of the system. The membranes used are eight inch diameter and sixty inches long thin film composite spiral wound membranes.

A clean in place (CIP) cleaning system is designed to enable cleaning of each stack individually. Argo Scientific Bioclean 103A and Bioclean 511 would be used to remove organic and inorganic foulant from the membranes.

5 CONCLUSIONS

The project posed a number of technical and logistical challenges during the design, procurement, installation and implement stages of the recycling scheme. The fast track project has culminated in the first in-building greywater recycling installation in the UK. The venue provides an ideal location to carry out on-going academic research into water recycling, water efficiency and conservation and to communicate the water wise message to a wide audience. Research is being carried out to assess the role of membrane processes as suitable technologies for future recycling of greywater and other alternative water sources in the 21st century.

References

1. The New Millennium Experience Company Ltd, *Annual Report and Financial Statements for the period ending 31 March 1998*.
2. CIBSE, *Chartered Institution of Building Services Engineers Guide*, Volume B, CIBSE, London, 1986.
3. D. Butler, E. Friedler and K. Gatt, *Wat. Sci. Tech.*, 1995, **31(7)**, 13.
4. J.B. Rose *Wat. Res.*, 1991, **25(1)** 37.
5. D. Christova-Boal, R.E. Eden, and S. McFarlane, *Desalination*, 1996, **106,** 391.
6. J. Murrer, and G. Bateman, *IChemE Conference, London, 26th March 1998*.
7. R. Birks, MSc Thesis, Cranfield University, School of Water Sciences, 1998.
8. J. Crook, *Wat. Sci. Tech.*, 1991, **24(9)**, 109.
9. N. M. Kayaalp, *Desalination*, 1996, **106,** 317.
10. J-H. Tay and P-C. Chui, *Wat. Sci. Tech.*, 1991, **24(9)** 153
11. B. Jefferson, *Wet News*, 15 Sept., 1998.
12. Gromaire-Mertz et. al., *Wat. Sci. Tech.*, 1999, **39,** 1.
13 D.F Samuel and M.A. Osman, *Chemistry of Water Treatment*, Butterworth,1983.
14. P.F. Cooper, G.D. Job, M.B. Green and R.B.E. Shutes, *Reed Beds and Constructed Wetlands for Wastewater Treatment*, WRC Swindon, Wiltshire, UK, 1996.
15. Building Services Research and Information Association (BSRIA) Report, 1997.

WASTEWATER RECLAMATION CASE STUDIES, THE BENEFITS OF OUTSOURCED MEMBRANE SYSTEMS

David Threlfall

Ecolochem International, inc.
Hydrohouse
Peterborough PE 2 6SE
United Kingdom

1 INTRODUCTION

Industrial and Domestic water consumption has increased significantly year on year Worldwide. As consumption has grown the costs of treated water have grown also. Municipal sewage is a low cost alternative that has become viable through the development of membrane water treatment systems. In the UK the recent shortages in domestic water supply have raised public awareness of recycling and reuse schemes.

This paper will present three examples of outsourced membranes systems used for municipal sewage reclamation. All these systems have been delivered on a Design Build Own Operate Maintain (DBOOM) contact by Ecolochem inc. in the NAFTA region and Ecolochem International, inc. in the UK. Operational performance and experience of these systems is presented for each system.

Outsourced membrane systems offer significant advantages to the end user; minimal or zero capital investment, reliability of supply, and 'hands off' system operation. The end user obtains a low cost alternative water supply without the concerns of manning, cleaning and maintenance that a capital system would require. Short or long term trial systems, without capital expenditure, at sizeable flowrates can be used to judge the economic advantage of outsourcing.

Treated sewage presents a serious challenge to the operation of an industrial water treatment system. The wide variety of organic and inorganic contaminants in the water process selection critical. Treated sewage will contain high levels of suspended solids, organic and inorganic colloids and biological material compared to usual well or surface water supplies. Reverse osmosis treatment of the supply will not only reject ionic species but also form a physical barrier to colloidal, bacterial and organic matter. The choices for reverse osmosis treatment are presented along with comparisons to other membrane techniques.

Table 1 - *Water Quality for Mexico site.*

	Raw feed	Treated water
TDS @ mg/l	600 ~1000	20 ~ 60
Silica @ Si mg/l	15 ~ 30	1 ~ 3
TOC @ C mg/l	5 ~ 20	<1
pH	6.4 ~ 7.5	6.5
TSS @ mg/l	1 ~ 20	< 1

2 OUTSOURCING EXAMPLES

Three industrial examples are presented, two in the Americas and one in the UK. In each case the system has been designed to provide a reliable supply of the highest quality to the end user.

2.1 Mexico

A Petrochemical refinery had access to a river supply for normal operating requirements. In times of drought the abstraction from the river was restricted and plant production was lost. The refinery had used treated effluent as cooling tower make-up water for twenty years, attempts to use this water for the ion exchange demineraliser feed water were unsuccessful. Non-ionic contaminants such as colloids, TOC and biological organisms were not removed in the treatment system and adversely affected the plant's boilers.

A 45 m^3/hr trial system was installed during 1993, consisting of two stage multimedia filtration followed by reverse osmosis and forced draft aeration. This system was proven to be reliable and subsequently emergency and long systems have been used for flow rates of 180 to 360 m^3/hr. currently a 360 m^3/hr system is installed supplying the refinery with boiler feed water from the municipal effluent. This system has operated for over four years producing 5 million m^3 of treated water with only a single set of membrane cleans. Table 1 shows the feed and product quality for the system. As the table shows the water produced from the system is of high quality.

The wastewater from the municipal works is chemically dosed with biocide and coagulant before the two pass filtration. The filtered water is then dosed with acid before cellulose acetate (CA) reverse osmosis. The acid dosing is used to prevent scaling and reduce hydrolysis of the CA membrane.

2.2 California, USA

A greenfield power station development, designed to meet peaking demands and reduce energy imports, needed high purity demineralised water for NO$_X$ control from a combined cycle gas turbine. The available raw water source was treated municipal waste.

The plant had to be designed to produce demineralised water for the plant at short notice and with intermittent operation, often with extended idle times. A two pass filtration system followed by two pass reverse osmosis was used to produce a high purity permeate. The final polishing for the NO$_X$ suppression injection is by a mobile deionisation unit. This trailer mounted single use demineraliser system can be exchange on exhaustion with an identical regenerated trailer, while the exhausted system is regenerated at a remote service centre. The plant was designed to produce 68 m^3/hr of demineralised water.

Table 2 - *Water Quality for California site.*

	Raw water	Second pass permeate
TDS @ mg/l	650 ~ 1000	< 7
Silica @ Si mg/l	34 ~ 65	< 0.2
TOC @ C mg/l	10 ~25	< 0.4
pH	6.8 ~ 7.5	6.5 ~ 7.5
TSS @ mg/l	5 ~ 50	< 1

The two pass RO system is CA/PA permeate staged, this give an overall rejection of 99% of ions reducing the load on the demineralisation system significantly compared to a single pass system. The plant operates at 55 ~ 60% recovery overall, the reject and filter backwash water is returned to the sewage treatment works.

The wastewater feed to the plant had high levels of turbidity, peaking at 100 NTU and returning an average of >40 NTU during operation. This was outside the design conditions and presented a severe challenge to the filtration system. Biweekly cleaning of the first pass membranes was required until the STW improved the quality of the effluent, when the turbidity fell to the design level the cleaning frequency fell to once per quarter. The demineraliser polishing unit is replaced on exhaustion, 15,000 ~ 30,000 m^3 of DI water are produced per exchange. The demineralised water has a consistent quality of 0.06 μScm^{-1}, < 20 ppb Silica.

2.3 United Kingdom

A smelting plant in the west of England needed to replace an ageing softener system, treating potable water for process steam. As an alternative to capital purchase of a new softener plant for the potable water supply the company approached a service company to provide an outsourced softening system. The site already had access to a supply of treated wastewater from the local municipal works; this water was used for washing and cooling. To reduce the costs of the softened water the service company proposed treating the treated sewage to the required specification.

Two pass filtration followed by reverse osmosis and forced draft aeration was chosen to give 105 m^3/hr of treated water. The local branch of the Environment Agency was concerned about the formation of disinfection by products (DBPs) from the use of sodium hypochlorite in the system. Other nonchlorinating biocides were assessed using laboratory jar tests. Performance was assessed for required dose rate, ease of use and effectiveness combined with filtration. By using a service company the customer did not have to worry about issues of membrane compatibility and equipment selection.

This site is commissioning for full production on 30th September 99.

3 SYSTEM DESIGN

All the outsourced systems presented have several design features in common. Reclamation of wastewater presents several challenges to any treatment system. All the presented designs have filtration and reverse osmosis (CA membranes)

3.1 Filtration

Reverse osmosis membranes are sensitive to suspended solids, using the SDI measurement gives a good indication of the suitability of the feed water for reverse osmosis. The two pass filtration system with in-line coagulant dosing produces an $SDI_{15} < 5$, this is suitable for CA feed water. Multi-media filtration is simple to use and at low to medium levels of suspended solids produces high quality effluent.

3.2 Reverse Osmosis

Reverse osmosis has been used widely for industrial water treatment as a pretreatment stage before ion exchange systems. The use of reverse osmosis will reduce the ionic load by 90 - 99%, and act as a barrier to particles, colloids and TOC. Cellulose acetate membranes are chosen over Polyamide (PA) or thin film composite (TFC).

Cellulose acetate membranes have a smooth uncharged surface, and are tolerant of free oxidants. This gives the membranes a very low fouling tendency. PA membranes have a rough surface and an anionic surface charge, these two factors coupled with a very low tolerance (1000 ppm.hours) to oxidants, and make PA membranes unsuitable to wastewater treatment, as they will foul very rapidly. The drawback with CA membranes is a susceptibility to hydrolysis. CA membranes need to be operated at a slightly acidic pH, however the use of acid reduces the Langlier Scaling Index (LSI) to <1 to produce a non-scaling environment.

3.3 Forced Draft Aeration

The use of acid feed to prevent hydrolysis in CA reverse osmosis lowers the pH and converts alkalinity to free carbon dioxide. The reverse osmosis membrane is permeable to dissolved gases, they have no charge and very low molecular weight, so the dissolved CO_2 is found at a very high concentration in the permeate. Forced draft aeration is used to reduce the CO_2 levels to < 10 ppm.

3.4 Alternative Systems

The development of ultrafiltration, microfiltration and membrane bioreactors are opening up other methods of wastewater treatment and reclamation. The systems described above utilise simple well understood technologies with an extensive track record. Currently combinations of MF or UF technologies with reverse osmosis (PA) can be used as an alternative to media filtration / (CA) RO. Micro and ultra-filters can be used to reduce the colloidal, particulate and bacterial contamination of the wastewater before polishing RO is used to reduce the ionic content. The membrane filters remove the common foulants from the feed water but leave dissolved nutrients. Both filtration systems are available in crossflow and dead end filtration modes, however they are liable to fouling from the rejected material and need frequent cleaning. The downstream PA membranes also foul relatively rapidly, over several days, as the nutrient content of the water feeds any biological growth in the system. Every membrane cleaning reduces the membrane life, and leads to increased costs over the lifetime of the plant.

Membrane bioreactors produce high quality effluent, but operate at very low flux, this makes the system expensive, as a very large membrane area is required. As yet there is insufficient capacity, in the right areas, to feed industrial needs. As the costs for the MBR systems fall then installations may become more attractive.

Other developments in the composition of polyamide membranes to give a 'low fouling' membrane are of interest. However until the membranes are resistant to free oxidants, such as chlorine, they are still likely to fouling several times faster than cellulose acetate.

4 BENEFITS OF OUTSOURCED MEMBRANE SYSTEMS

An outsourced system is designed to meet the customer's demands for quantity and quality of treated water. To meet this end through the service contract and to minimise the operating costs the design will be conservative and based on reliable technologies. Competition between capital equipment manufacturers is intense and design integrity may be reduced to meet a cost.

Common techniques to lower the capital coat for equipment is to increase flux reducing the number of membranes, to specify low pressure polyamide membranes or fit high surface area membranes. Each of these techniques will lower the cost of the system but can lead to high operating costs and busy maintenance schedules. All these methods lead to high fouling rates, this may not be observed during the short commissioning period but will soon be apparent. Frequent cleaning of membranes is expensive and gives the added burdens of extra waste disposal and chemical handling. Without a flexible design, responding to changes in the feed water quality may be difficult.

An outsourcing contract is typically for 5 to 15 years, over this time the service company has to meet the contract specifications or payment may not be received. Long term operational experience with a wide variety of systems gives the service provider a pool of knowledge to draw on without resorting to expenses external consultants. An additional benefit is the available provision of mobile emergency equipment, this can be used in the event of the unexpected or to increase the supply of treated water over short periods or to cover plant outages.

The removal of operating cost from the customer by the service company long term economic plans can be drawn up to compare the anticipated costs between capital or outsourced systems. The 'cost or ownership' for outsourced systems has been shown to be lower than capital purchase over the lifetime of the contract.

At anytime during the contract the service provider will offer the opportunity to increase efficiency if technology is available. This option is available without additional expenditure, unlike capital purchase, and may reduce costs significantly.

5 CONCLUSIONS

Industrial feed water production from reclaimed wastewater will play a larger role in the future. Industry has a great demand for water but is seen as a lower priority than domestic or agricultural users. The availability of treated sewage for exploitation is limited by the provision of pipelines to potential industrial users. Why put water to river discharge, at a cost, that could be sold as a commodity?

The case histories presented have been chosen to illustrate the situations where outsourcing can delivery a low risk, economic advantage over a capital system. The choices made by the service company differ from those of the equipment supplier, and are made on the basis of long term objectives.

Outsourcing water treatment services is an increasingly common choice. The benefits of fixed costs, and guaranteed production over the lifetime of the contract are advantages that capital purchase cannot meet.

References

R. T. Taylor, in *Proceedings of the 58th Annual Meeting, International Water Conference,* Engineers' Society of Western Pennsylvania, 1997, p 442.

S. Morris, *Water and Environment Manager*, 1999, March, p10

B. Birkenhead, 'Wastewater is now a viable economic water source for water companies and industry', presented at Eurochem/ET, June 3rd, 1998, Birmingham.

COMPARISON BETWEEN DIFFERENT OUT-TO-IN FILTRATION MF/UF MEMBRANES FOR THE RE-USE OF BIOLOGICALLY TREATED WASTEWATER EFFLUENT

Emmanuel Van Houtte, Johan Verbauwhede, Frans Vanlerberghe and Johan Cabooter

Intermunicipal Water Company of Veurne-Ambacht (I.W.V.A.),
Doornpanne 1,
B-8670 Koksijde,
BELGIUM

1 INTRODUCTION

The Intermunicipal Water Company of Veurne-Ambacht (IWVA) and the Municipal Water Company of Knokke-Heist, producing and distributing drinking-water respectively at the western and eastern part of the Flemish coastal area, plan to re-use wastewater effluent for the artificial recharge of the existing dune water catchments. Artificial recharge would create a sustainable groundwater production preventing the ingress of salt water under the dunes.

Microfiltration was chosen to pre-treat the effluent; to remove salts and nutrients the produced filtrate would be further treated with reverse osmosis.

As a result of previous tests[1] using microfiltration and ultrafiltration membranes it was concluded that **out-to-in filtration** was the most effective way to treat wastewater effluent. It was decided to evaluate different systems with out-to-in filtration before tendering the full-scale plant. Three systems competed: two submerged systems (CMF-S from USF MEMCOR and ZeeWeed from ZENON) and one pressurised system (CMF from USF MEMCOR). All worked in dead-end mode.

This paper describes the results of these comparative tests.

2 DESCRIPTION OF TREATMENT SYSTEMS

2.1 Testing procedure and data acquisition

The tests were done from June 1st until October 16th 1999. Both the CMF and the ZeeWeed installation performed during the full period. Tests with CMF-S started on August 18th.

The pressure and flow were logged continuously. The quality of the produced filtrate was compared to the feed water (effluent) quality. The following parameters were controlled on a periodic base: acidity (pH), conductivity, BOD, COD, total phosphorous, nitrate, ammonia, sulphate, total coliforms, faecal coliforms, faecal streptococci, heterotrophic plate counts (HPC) and UV-absorption.

The SDI_{15} of all filtrates was controlled periodically and for the last month also particle counting was performed on the filtrate of the submerged systems.

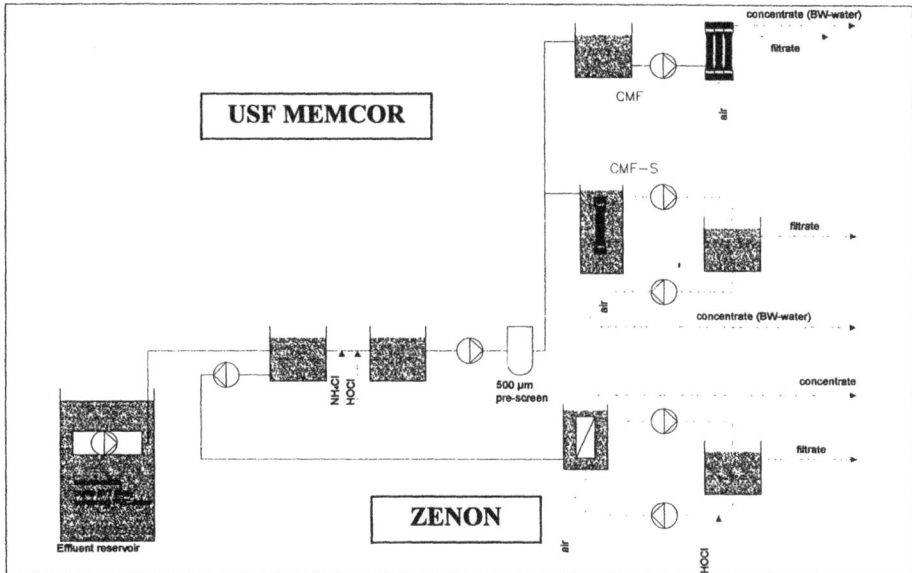

Figure 1 *Process scheme of comparative tests*

2.2 Description of the systems

2.2.1 Effluent intake and pre-treatment The raw effluent was taken from a reservoir with two submersible pumps both housed in a 1mm screened PVC-tube. This pre-filtered effluent entered a first $1m^3$ reservoir. From this first reservoir the feed was taken for the ZeeWeed installation.

The effluent flowed from the first to a second $1m^3$ reservoir, but in between ammonia and chlorine (a mean of 1.4 mg free chlorine/l) were dosed. From this second reservoir the chloraminated effluent was pumped through an automatic backwashing 500μm strainer to feed the CMF and CMF-S installations.

2.2.2 CMF installation It was a 3M10C unit from USF MEMCOR containing 3 modules; each housing 20,000 polypropylene (PP) hollow fibres with a nominal pore size of 0.2μm; the total membrane surface is 45 m².

The filtration, with a flux of 3,240 l/h, is from the shell (outside) to the lumen (inside) side of the fibre where the filtrate is collected. The contaminants accumulate at the surface of the membranes causing an increase of pressure with time. At the end of each filtration cycle, either at a preset time (20 minutes) or earlier if the rate of resistance to filtration is increasing rapidly, a gas backwash is initiated. This gas backwash – explosive release of compressed air from lumen to shell side of the fibre - dislodges the contaminants from the surface of the membrane after which they are removed from the system by a cross-flow of feed water.

As solids gradually accumulated over time a periodic chemical cleaning (CIP) was required when the transmembrane pressure (TMP) reached 130 kPa. A CIP consisted of an acid clean followed by a caustic clean and restored the TMP below 50 kPa.

2.2.3 CMF-S installation   It was a 1S10T unit from USF MEMCOR. The module is equipped with the same fibres as the CMF unit. One S10T module contains 13 m² of surface area.

Initially the filtration time was 15 minutes at a flux of 1,000 l/h; the clean water is drawn through the membrane wall by pump suction pressure into the centre of each fibre[2]. During the backwash low pressure air is forced through the bundle of the membrane module (from bottom to top) creating a scouring effect on the external fibre surface. A short filtrate flow (2,250 l/h) from the lumen to the shell side of the fibres supplements the air to further dislodge the contaminants that are drained from the tank at the end of the backwash cycle[2]. After the reservoir is refilled with feed water production can restart.

Chemical cleaning is achieved by contact of the cleaning solution with the membranes in combination with a periodic aeration. The CIP is performed when the TMP was 85 kPa and restored the TMP to 20 kPa.

On September 16th the flux was increased to 1170 l/h; on October 1st the filtration time was set to 20 minutes.

2.2.4 ZeeWeed installation   The ZeeWeed module (ZW-500) contained chlorine-tolerant polyvinylidene fluoride (PVDF) hollow fibres with a nominal pore size of 0.035µm and an absolute pore size of 0.1µm; the total surface area was 42.8 m².

The membranes operate under a partial vacuum created within the hollow fibres by the operation of a centrifugal pump[3]. A filtration period of 215 seconds at a flux of 1,800 l/h is followed by a back pulse of 15 seconds. During a back pulse low pressure air introduced at the bottom of the membrane bundle is combined with a reversed permeate flow (2,100 l/h). The contaminants are removed from the reservoir by a small overflow during this back pulse.

After a number of backpulses an extended back pulse is performed. Chlorinated filtrate flow (1,000 l/h) was pumped through the membranes for 120 seconds followed by a short soak period of 120 seconds, always combined with air.

Table 1. *Characteristics of all three out-to-in MF filtration systems used*

	CMF	CMF-S	ZeeWeed
Membrane fibre	0.2µm pore size PP, symmetric and hydrophobic non chlorine resistant	0.2µm pore size PP, symmetric and hydrophobic non chlorine resistant	0.1µm pore size PVDF, asymmetric and hydrophilic chlorine resistant
Driving force	Positive pressure via pumped feed	Suction pressure via centrifugal pump	Suction pressure via centrifugal pump
Backwash	Pressurised air from in-to-out and feed sweep alongside membranes	Low air pressure on outside of membranes filtrate from in-to-out and draindown	Low air pressure on outside of membranes filtrate from in-to-out with small overflow
CIP	Acid and caustic (4h)	Acid and caustic (4h)	Chlorine and hydrochloric acid (2h)
Biofouling prevention	Chloramination of feed water	Chloramination of feed water	Periodic back pulse with chlorinated filtrate

The absolute pressure in the system varied from 1.00 atmospheric bar to 0.45 atmospheric bar. At that stage a chemical cleaning is performed by emptying the reservoir and filling it with chlorinated drinking water or filtrate. The system soaks for 60 to 90 minutes alternated with short cycles of back pulses and filtrate production. At the end hydrochloric acid is added to remove scaling from the membranes; the last step is to drain the reservoir.

On September 9th the module was replaced. The new module had 46 m^2 of surface area. The flux was increased to 1840 l/h.

3 RESULTS OF COMPARATIVE TESTS

3.1 Hydraulic results

The ZeeWeed and CMF worked with constant fluxes respectively 40 and 72 l/h.m^2. The CMF-S initially had an initial flux of 77 l/h.m^2 but it was increased to 90 l/h.m^2.

3.1.1 CMF installation The mean recovery over the whole period was 87.5 % with a maximum of 95.4 %. The mean production period was 14 minutes. The frequency of CIP's diminished as the temperature of the feed water increased: in June a CIP had to be performed every 1 or 2 weeks but in July, August and September a CIP was only needed once in 3 or 4 weeks time. It should be mentioned that during the summer shutdowns of the CMF system occurred regularly due to clogging of the pump and the screened PVC tube in the effluent reservoir. After a shutdown the TMP of the CMF, but also of the CMF-S, was always lower compared with the TMP before the shutdown (figure 2). With a constant flux of 72 l/h.m^2 the daily effective filtrate production varied between 60 and 68 l/h.m^2.

3.1.2 CMF-S installation Because the BW-interval was constant the recovery over the whole period was constant: 88.7 %. The installation was only cleaned once during the 2 months of testing; the TMP increase was very slow.

Even when on the CMF the BW-interval decreased below 15 minutes, the CMF-S performed very well on the same feed, with a constant BW-interval of 20 minutes and with a higher flux. Generally the TMP increase was 5 kPa during a production period, but it was noticed that the backwash was able to restore the TMP when occasionally greater TMP increases occurred.

With a constant flux of 77 l/h.m^2 the daily effective filtrate production varied between 57 and 65 l/h.m^2; when the flux was increased to 90 l/h.m^2 it varied from 70 to 78 l/h.m^2. Due to problems with the feed pump between October 6th and October 12th, the feed for the CMF-S was changed to effluent that only passed the 1mm screened PVC-tube. This did not have a negative effect on the performance of CMF-S so it could be concluded that a 1000μm strainer would be sufficient on CMF-S.

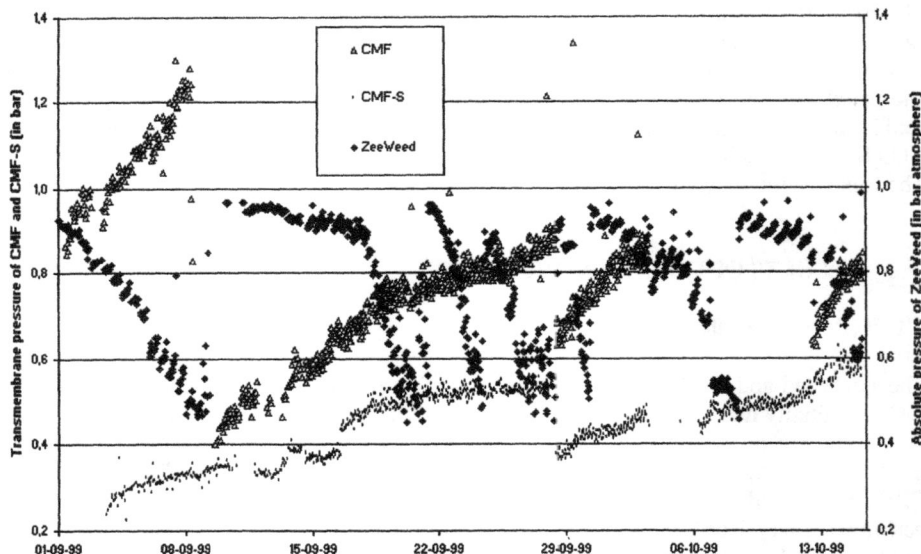

Figure 2. *Performance of the compared MF out-to-in systems*
from September 1st until October 15th

3.1.3 ZeeWeed installation It was not very easy to obtain constant recoveries, as the feed flux was not constant. After a flow controller was installed it proved possible to feed the ZeeWeed reservoir only while the system was in production mode; overflow only occurred during the back pulse. In the last runs - after the flow controller was installed - recoveries of 87.5 % ware achieved with a weekly CIP (figure 2). At that time the system already worked on raw effluent as one of the 1mm screened PVC tubes around a submersible pump was broken (September 28th) and no strainer was placed between this pump and the ZeeWeed reservoir.

After a chemical cleaning the pressure drop during a production period was 1 to 2 kPa. However as the pressure built up over time this TMP drop gradually increased (figure 2).

The mean daily effective filtrate production was between 30 and 34 $l/h.m^2$ with a mean effective production over the period June 1st –September 30th was 29 $l/h.m^2$; this includes CIP cycles and shutdowns in that period. Most of the shutdowns were due to a decline in feed flux before the flow controller was installed. Because no pre-screen was needed shutdowns due to clogging of the PVC-tube did not influence the ZeeWeed installation.

3.2 Quality

Filtrate samples were taken on a weekly base for bacteriological control. Before sampling chloramination had been stopped during some hours. The SDI_{15} was controlled regularly and during the last month of the test a particle counter monitored on line both the filtrate from CMF-S and ZeeWeed.

3.2.1 CMF installation The SDI_{15} had a mean value of 1.9 with a minimum of 0.9 and a maximum of 3.4. Generally the SDI's were lower in dry periods (no dilution of sewage with rainwater).

The organic load of the effluent was partially removed by CMF (table 2). No bacteria were found in the filtrate and the log removal for HPC was around 3.

Table 2. *Quality of the wastewater effluent of the WWTP at Wulpen compared to the filtrate quality of CMF and ZeeWeed*

Parameter	Effluent	CMF filtrate	ZeeWeed filtrate	Number of samples
Acidity (pH)	7.6	7.7	7.8	26
Total phosphorous (mg P/l)	0.52	0.40	0.37	8
UV 254 absorption (cm$^{-1)}$	0.2686	0.2426	0.2424	8
Total Organic Carbon (mg C/l)	24.2	18.5	18.4	8
Biological Oxygen Demand (mg O$_2$/l)	8	3	3	8
Chemical Oxygen Demand (mg O$_2$/l)	39	33	33	8
Total coliforms (counts/ml)	10^5-10^6	0	0	14
Faecal coliforms (counts/ml)	10^5	0	0	14
Faecal streptococci (counts/ml)	10^4-10^5	0	0	14
Log removal HPC 37°C (48 h)	10^4-10^5	3.15	2.83	14
Log removal HPC 22°C (72 h)	10^4-10^5	3.20	2.54	14

3.2.2. CMF-S installation The mean SDI$_{15}$-value was 2.8 with a minimum of 1.4 and a maximum of 4.3. However in the last weeks of the tests the SDI's improved and were comparable to those of the CMF and ZeeWeed filtrate.

The organic load of the effluent was partially removed; no bacteria were found in the filtrate (table 3). The log removal for HPC was around 2 but it the counts increased with time. This could be due to the location of the installation. Unlike the CMF, both the CMF-S and the ZeeWeed installations were installed in open air near the aeration basin of the sewage plant. Small sewage droplets in the atmosphere could have contaminated the air (compressor) and the filtrate reservoir. A sample taken while chloraminating the feed resulted in lower counts that proved the effectiveness of this method.

3.2.3. ZeeWeed installation In the first part of the tests the mean SDI$_{15}$-value was 3.3 but after replacement of the module it decreased to 2.6 with a minimum of 1.6 and a maximum of 3.5.

The organic load of the effluent was partially removed and no bacteria were found in the filtrate (table 2). The mean log removal for HPC was around 3 but here also a small increase of counts was observed although not in the same extent as on the CMF-S.

The pH of the filtrate produced with the ZeeWeed unit was slightly higher (table 2) presumably by stripping of carbon dioxide during the back pulse[4] (airflow).

3.2.4. Particle counting On September 8th a Met One 215W particle counter was installed on both submerged systems. It continuously counted particles greater than 1µm. Due to the backwash it was difficult to get stable counts, especially on the CMF-S system where the backwash cycle lasted longer.

On the ZeeWeed filtrate, the amount of particles never exceeded 20 counts/ml and most of the results were between 5 and 10 counts/ml; on the CMF-S filtrate it was slightly higher with a majority of counts between 0 and 20 counts/ml.

Table 3. *Quality of the wastewater effluent of the WWTP at Wulpen compared to the filtrate quality of CMF-S*

Parameter	Effluent	CMF-S filtrate	Number of samples
Acidity (pH)	7.6	7.7	7
Total phosphorous (mg P/l)	0.57	0.37	3
UV 254 absorption (cm$^{-1)}$	0.3119	0.2812	3
Total Organic Carbon (mg C/l)	28.3	21.7	3
Biological Oxygen Demand (mg O$_2$/l)	8	4	3
Chemical Oxygen Demand (mg O$_2$/l)	39	35	3
Total coliforms (counts/ml)	10^5-10^6	0	7
Faecal coliforms (counts/ml)	10^5	0	7
Faecal streptococci (counts/ml)	10^4-10^5	0	7
Log removal HPC 37°C (48 h)	10^4-10^5	1.87	9
Log removal HPC 22°C (72 h)	10^4-10^5	1.93	9

4 CONCLUSION

Three out-to-in microfiltration systems were compared with one another. ZeeWeed and CMF-S were submerged systems; the CMF is a pressurised system.

All three systems reached recoveries over 85 % and could produce for at least one week without the need to be chemically cleaned. The filtrate produced with all systems was comparable and free of bacteria.

All MF systems that were tested proved capable of producing a filtrate quality ready to feed a reverse osmosis installation.

The submerged systems showed some advantages over the pressurised system. The footprints are smaller and the concept is less complicated compared to the pressurised air backwash of CMF; a 500μm strainer is no longer needed so coarser strainers could be sufficient. This could mean that investment and maintenance costs would be smaller. As the transmembrane pressures are smaller the initial pressure could be restored very easily. A disadvantage is that submerged systems use filtrate for the backwash that means that the risk of recontamination of those systems is higher. Care should be taken to prevent recontamination of the filtrate. However chloramination not only proved to enhance the performance of CMF[5] but also showed to be an efficient way to prevent recontamination of the filtrate.

Using chlorine resistant membranes is an advantage as the risk on damage is smaller and the chlorine and ammonia consumption for bio-fouling prevention could be reduced. However the hydrophobic PP membranes seem to have a slower TMP increase which means that chemical cleanings should be performed less frequently.

Standardisation of the submerged modules, which would make them interchangeable, could in the future benefit to the customers, as the investment risks would be reduced.

ACKNOWLEDGEMENT

AQUAFIN gave the opportunity to do the tests on the wastewater effluent at Wulpen WWTP and actively supported the IWVA in the project. It is important to mention the support from USF MEMCOR and ZENON. Their collaboration made that the trial work could be done under good conditions.
We want to thank all the people from IWVA and Knokke-Heist who helped us with the tests especially Delphine, Luc P., Luc C., Pol and Sven.

References

1. E. Van Houtte, J. Verbauwhede, F. Vanlerberghe, S. Demunter and J. Cabooter, *Treating different types of raw water with micro- and ultrafiltration for further desalination using reverse osmosis.* Proceedings 'Membranes in drinking and industrial water production' Conference, Amsterdam, Desalination 117, p 49-60, 1998.

2. B. Birkenhead, *Microporous separation technologies : principles and economics.* Proceedings Workshop on 'Microporous membrane filtration technology improving drinking water quality', Genoa, 1999.

3. D. Mourato, G. Best, M. Singh and S. Basu, *Reduction of disinfection byproducts, dissolved organic carbon and colour using immersed ultrafiltration membranes.* Proceedings AWWA 'Membrane Technology' Conference, Long Beach, 1999.

4. G. Leslie, M. Patel, J. Norman, W. Dunivin, E. Martin and R. Sudak, *Microporous membrane pre-treatment options for reverse osmosis in municipal wastewater reclamation applications.* Proceedings AWWA 'Membrane Technology' Conference, Long Beach, 1999.

5. E. Van Houtte, *Re-use of wastewater effluent for artificial recharge in the Flemish dunes.* Proceedings Workshop on 'Microporous membrane filtration technology improving drinking water quality', Genoa, 1999.

Industrial Applications

SULPHATE REMOVAL MEMBRANE TECHNOLOGY : APPLICATION TO THE JANICE FIELD

G.H.Mellor[1], R.C.W.Weston[1], G.F.Bavister[1], and A.White[2]

[1]Consept
Albany Park Estate
Camberley GU15 2QQ

[2]Kerr-McGee North Sea (UK) Ltd
Ninian House
Crawpeel Road
Aberdeen AB12 3LG

1 ABSTRACT

The Janice Field is a small marginal field, which has recently started production. The field is being developed using a Floating Production Vessel (FPV) and subsea production and injection wells. Water injection is required in order to maintain reservoir pressure and improve reservoir sweep. The formation water contains approximately 10,000 mg/l Ca, so with seawater containing approximately 2900 mg/l sulphate, severe calcium sulphate scaling was predicted to occur where ever seawater and formation water mix.

Seawater injection would result in increased operating costs, as a result of scale squeeze chemicals and the increased frequency of remedial workovers. This increased cost was greater than that to install and operate a Sulphate Reduction Package and inject Low Sulphate seawater. The decision was taken to install the world's first floating Sulphate Reduction Package on the Janice A FPV.

This paper presents some the reasons why Low Sulphate Seawater injection and sulphate removal membrane technology were selected and later describes the design of the system and its integration on the FPV.

The Janice A sulphate reduction package was commissioned during August, 1999. To date the package has consistently produced low sulphate water of quality exceeding the design specification requirements.

Kerr-McGee North Sea (UK) Ltd is a wholly owned subsidiary of Kerr-McGee Corp., an Oklahoma City-based energy and chemical company with assets of $5.6 billion. Kerr-McGee North Sea (UK) Ltd are the field operators.

Consept are process engineering consultants specialising in membrane separation technology applications.

Axsia Serck Baker are a licensed original equipment manufacturer for the application of this technology, which is patented by Marathon Oil and sub-licensed by Dow.

2   FIELD OVERVIEW

The Janice Field is located some 275 kms East-Southeast of Aberdeen on the south western margin of the West Central Graben in North Sea Block 30/17a .

Kerr McGee North Sea (UK) Limited was appointed operator of the block in March 1996.

The field received project sanction in September 1997, with first oil being achieved some 17 months later on 9th February 1999.

3   GENERAL FACILITIES OVERVIEW

The field is being development using the Janice A Floating Production Vessel (FPV), a converted accommodation vessel (formerly the West Royal). The production and water injection wells are drilled from a single subsea cluster some 450 m away from the Janice A. The Janice A FPV is equipped with a single separation train, a test separator, water injection and gas lift facilities. Crude oil is exported via a 14" pipeline to the Teeside via the Phillips operated Norpipe line. Gas is exported via a 12" pipeline to the Phillips operated Judy platform.

4   FORMATION WATER CHEMISTRY

The need for water injection was identified very early on during the development stage for the Janice Field. Water injection is need for reservoir pressure maintenance and to sweep the oil towards the production wells.

The most readily available and obvious source of injection water is filtered seawater. However, down hole formation water samples from Janice were recovered, and found by two independent analyses to contain 10,000 mg/l Ca. With seawater containing 2900 mg/l Sulphate (SO_4), this gives rise to the deposition of anhydrite ($CaSO_4$) wherever the waters mix, in almost any ratio. The most severe potential consequence is seawater breakthrough at the production wells, where anhydrite deposition could result in formation damage in the near wellbore area, as well as scale deposition inside the liner and tubing.

Scale prediction calculations were run using the Oklahoma University program OKSCALE. A maximum of 1300 mg scale was expected at 70% seawater breakthrough. Anhydrite deposition is less severe, but still significant, at almost any mixing ratio.

To put these values in context other operators have reported serious scaling problems when only a couple of hundred mg/l $CaSO_4$ is predicted to deposit.

	Formation Water * (mg/l)	Seawater (mg/l)	Low sulphate sea water, LSSW (mg/l)
Sodium	55810	11500	10530
Potassium	2470	397	248
Calcium	10170	410	131
Magnesium	755	1290	526
Chloride	102410	20454	18170
Sulphate	255	2780	62
Bicarbonate	650	134	41
Total	172520	36970	29708

Table 1 *Janice formation water compositions and that of filtered seawater.*

5 ALTERNATIVE WATER SOURCES

Injecting aquifer water from an overlying reservoir for pressure maintenance and sweep was considered. The aquifer water is compatible with the formation water so avoiding any scaling problems. Unfortunately, neither an overlying nor underlying aquifer exists near the Janice Field, which rules this alternative water source out.

Re-injecting produced water for pressure maintenance and sweep was another possibility but was discounted for a number of reasons:
□ delay in water production from wells causes reservoir pressure to fall and loss of reserves
□ possible increased level of reservoir souring
□ possible asphaltene carry over in the produced water and subsequent injectivity problems
□ possible higher solids loading in produced water and subsequent injectivity problems
□ possible fouling by iron salts and by increased levels of biological activity
□ reduced degree of thermally induced fracturing
This leaves only two other real possibilities, namely:
□ Filtered seawater, with scale inhibitor squeeze treatments
□ Low Sulphate Sea Water (LSSW).
The only real advantage for filtered seawater is that it is readily available. There are however, a large number of disadvantages. A few are listed below:
□ Increased chemical cost for scale inhibitor squeeze treatments
□ Potential impairment of wells due to squeeze treatments (loss of productivity/ relative permeability effects / water blocking)
□ Production deferment during squeeze treatment and soak period
□ Requirement for analysis, logging and monitoring to estimate treatment effectiveness (difficult in subsea horizontal wells).
□ Difficulty of adequately designing squeeze chemical placement, especially in long (horizontal) completion intervals potentially compounded by reservoir heterogeneity and cross flow.
□ Special requirements for well design to optimise squeeze placement, e.g. blank sections in screen assemblies

- In the event of squeeze treatments being inadequate, one or more of the following well interventions may be required: (1) frequent repetition of squeeze; (2) re-perforation; (3) solvent washes; (4) milling with coiled tubing; (5) pulling/replacing tubing; (6) side-tracking/re-drilling.
- Other Operators experience at scale squeezing horizontal wells discouraging
- The main advantage of Low Sulphate Seawater is that it is compatible with the formation water.
  There are however, also a number of disadvantages. A few are listed below:
- Increase Capex
- Increased Space (especially where space is limited)
- Weight (especially where weight is critical)
- Increased Opex in connection with maintenance of the membrane package and ancillary equipment.
- Increase filtration
  Well intervention forecasts were made for the proposed Janice field development options with and without a Sulphate Reduction Package (SRP) to produce LSSW. Even assuming wells can be scale squeezed satisfactorily, the cost of the increased frequency of remedial workovers, if seawater were used far exceeded the cost of installing and operating a SRP unit. It was therefore recommended that a SRP unit be used in Janice for the provision of LSSW.

6   SULPHATE REDUCTION PROCESS

The process is based on *nanofiltration* membrane separation. *Nanofiltration* is a membrane process that selectively removes sulphate ions to produce a reduced sulphate sea water. The process is similar to reverse osmosis which is used extensively worldwide for desalination of seawater, however, nanofiltration has a better feed to permeate conversion at 75%. The concept has been technically proven by its successful application in both North Sea and Gulf of Mexico water injection plant [1,2,3] with an operating capacity in excess of 330 mbd (53Mld). The process utilises nanofiltration membrane supplied by Dow as described next.

7   DOW FILMTEC SR90-400 MEMBRANE

The SR90-400 membrane is defined as a thin film composite membrane consisting of three layers: a polyester support web, a microporous polysulfone interlayer, and an ultra thin barrier layer on the top surface.

In the SR90-400 membrane the barrier has a negative surface charge which repels anions (negatively charged ions). Monovalent anion rejection (for example chloride ions) decreases with increasing salinity because the high concentration of positive cations starts to shield the negative surface charge at the membrane surface. Divalent ions (for example sulphate) have a high charge density such the shielding effect is negated and these ions are then preferentially rejected by the membrane barrier layer.

The membrane is constructed into a spiral wound element, and a system comprises of several spiral wound membrane elements installed inside a pressure vessel. Pressurised water flows into the vessel and through the channels between the spiral windings of the element. Elements are connected together in series within a pressure vessel. The feedwater becomes more and more concentrated and will enter the next element, and at last exits from the last element to the concentrate valve where the applied pressure will be released. The permeate of each element will be collected in the common permeate tube installed in the centre of each spiral wound element and flows to a permeate collecting pipe outside of the pressure vessel.

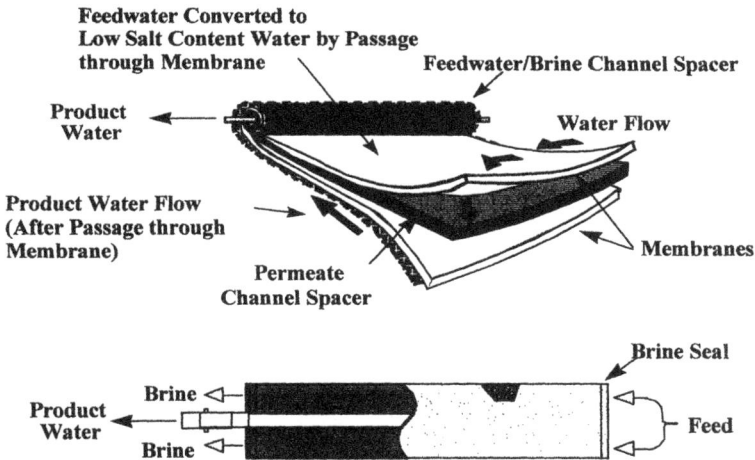

Figure 1 *Detail of spiral wound thin film composite membrane module*

In practice the membrane vessels are normally arranged in two stages with an approximate 2:1 staging ratio (ie twice the number of membrane vessels in stage 1 to stage 2). The concentrated reject water from the 1st stage membrane vessels is supplied as feedwater to the 2nd stage vessels where a further stage of treatment occurs. The low sulphate seawater (lssw) product water from the two stages is then collected to form the overall lssw product stream. With this system it is possible to increase the % recovery (or water conversion factor) which minimises the weight and size of the package.

8 NANOFILTRATION OPERATING CONDITIONS.

Proper pre-treatment of the feedwater supplied to the nanofiltration membrane is required to maximise the efficiency and life of the membrane elements; and to ensure trouble-free operation. Pre-treatment requirements include:
☐ Removal of fine suspended solids which can plug or block the membrane surface.
☐ Prevention of biological growth on the membrane surface

❑ Prevention of scale formation on the membrane surface during concentration of the feed.

❑ Removal of any oxidising biocides (eg chlorine) which can damage the membrane.

❑ Pressurisation as required to achieve nanofiltration separation.

The sulphate reduction package for Janice is designed to achieve all of the above objectives.

9   PROCESS CONSTRAINTS

The membranes must be supplied with seawater with a Silt Density Index (SDI) measured at less than 5 (note: SDI is a measure of the presence of all colloidal particles greater than 0.45 micron). There is no direct correlation between this index and, say, particle removal efficiency except that, generally, the better the particulate removal the lower the SDI.

To enhance the particle removal efficiency of the multi media filters in order to achieve SDI's below 5, the presence of residual chlorine is recommended. The residual chlorine remaining after filtration must itself be removed since it substitutes itself with radical groups in the membrane polymeric structure and over a short time irrevocably damages the membrane.

Since the membrane possesses a negative surface charge, cationic polyelectrolytes cannot be used as filtration aids since there always exists a risk of polyelectrolyte breakthrough resulting in serious fouling on the membrane surface.

Positive removal of chlorine is achieved by dosing an excess of a suitable de-chlorination chemical (commonly oxygen scavenger is used in offshore applications). As salts present in the feed are concentrated in the reject streams from the membrane system, antiscalant is injected upstream of the cartridge filters to inhibit the formation of scale. Feed water then passes through a cartridge filter to trap any particulate matter that may be present due to carryover from the multi media filters.

The feed pressure is then boosted to enable nanofiltration to occur and provide sufficient product pressure to enable entry into the downstream vacuum de-aerator tower.

The overall system operates at a conversion of 75% where conversion is defined as the percentage ratio of product from the membrane system divided by feed to that system, e.g. for very 100 m^3/h of seawater feed, 75 m^3/h of low sulphate water will be produced.

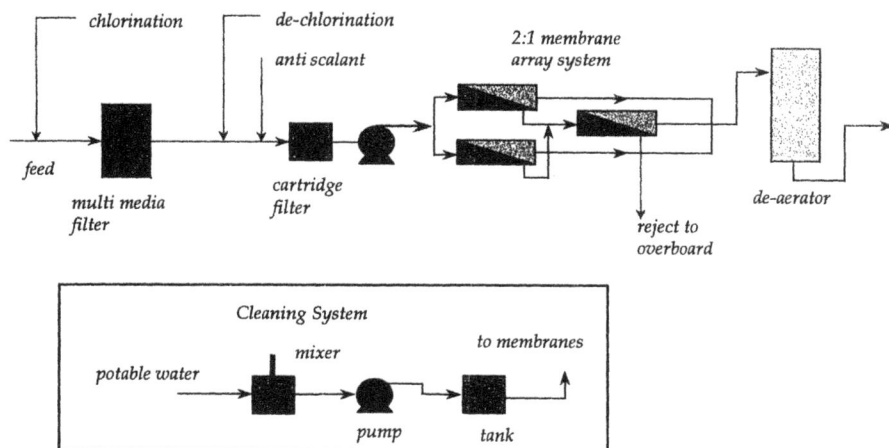

Figure 2 *Typical process flow diagram for sulphate reduction process*

10 SULPHATE REDUCTION PACKAGE EQUIPMENT SCOPE

The sulphate reduction (SR) package comprises of the following major items installed on a multiple deck structure designed for a single lift:

- ❑ Multi media filters (3 x 33 %, or 2 x 50% with one in backwash)
- ❑ Airscour blowers (2 x 100 %)
- ❑ Chemical injection points for antiscalent and oxygen scavenger chemicals
- ❑ Spare chemical injection point for on-line biocide dosing (if required)
- ❑ Static mixer (1 x 100 %)
- ❑ Cartridge filters (2 x 50 %).
- ❑ SR booster pumpsets (2 x 50 %)
- ❑ SR membrane pressure vessels (2 x 50 % trains)
- ❑ SR membrane elements (2 x 50 % trains)
- ❑ Clean in place (CIP) equipment c/w CIP tank; CIP pump; CIP cartridge filter.
- ❑ Local package control panel c/w PLC and HMI unit
- ❑ Field and panel mounted instrumentation
- ❑ Interconnecting piping; fittings and supports.
- ❑ Manual and actuated valves
- ❑ Control valves
- ❑ Interconnecting instrument and control cabling; traywork and supports
- ❑ Supporting structure

11 INTEGRATION WITHIN THE JANICE WATER INJECTION SYSTEM

The SRP package is integrated within the Janice water injection system as follows:

- Two seawater/firewater pumps (located in the aft inner port and starboard) columns supply seawater.
- Coarse straining to 100 microns
- Hypochlorite generation package
- Seawater is heated in the glycol/water cooling medium heat exchangers
- The seawater stream is then split with 93333 bpd supplied to SRP, and 10000 bpd bypass
- SRP package is located on 2nd deck central
- Vacuum dearation
- High pressure water injection.

12 CONTROL & INSTRUMENTATION

The package control system has been designed for automatic operation with limited operator intervention.

The normal operating mode is for semi-automatic startup sequences, and in this mode the operator initiates a sequence which then proceeds automatically through a number of stages. The following sequence only proceeds with further operator intervention. Automatic shutdowns occur on fault.

Each sulphate removal train is designed to operate at 75% conversion at design output. There is no turndown during normal operation (excess water not required for injection can be dumped overboard). If the inlet seawater temperature falls below the design range then the train output is automatically reduced. The level of turndown is restricted to ensure that the SR membranes are always operated within specified flow limits.

Each SR train start-up sequence includes a pre-flush followed by automatic booster pump start-up. Low sulphate product water is initially discharged overboard for a time period to allow for performance stabilisation. Once the quality is proven then the LSSW can be forwarded to the de-aerator.

An offline sequence is provided for SR membrane cleaning in place (CIP) operations. The operator is required to prepare the correct cleaning solutions for this procedure and to position the valves and operate the CIP pump manually.

13 DATA LOGGING

On-line performance data (e.g. flowrate, temperature, pressure, conductivity) is communicated to the central control facility for data logging purposes.

This data set is then sent to Axsia Serck Baker for monitoring and normalisation. The normalisation process shows how the membrane plant is performing against the base line conditions, enabling the prediction of manual intervention, e.g. to plan when to clean the membranes.

In addition to the on-line performance monitoring data, a daily program of manual sampling and test procedures (including silt density index (SDI) and sulphate tests) are performed by the laboratory technicians and operator.

14 SULPHATE REDUCTION PACKAGE PROJECT MILESTONES

The contract was awarded to Axsia Serck Baker on a fast-track basis in June 1997. Due to the central location of this package on the vessel it was critical to the overall vessel construction program that the package was delivered on time. The completed package was delivered on-time and to budget during February 1998.

The completed vessel float-out occurred during November 1998 and first oil was produced during February 1999. The sulphate removal package was commissioned during August 1999 and has consistently produced low sulphate water of quality exceeding the design specification requirements.

15 CONCLUSIONS

Use of a Sulphate Reduction Plant (SRP) on the Janice field is a cost effective alternative for seawater injection resulting in a lower overall field operating expense.

Sulphate Removal membrane technology is now a proven technology, delivering very high quality injection water.

The field development can be optimised with horizontal wells avoiding the need to scale squeeze.

Increased production up time is achieved by avoiding the deferral of production associated with scale squeeze treatments.

Low-Sulphate Seawater (LSSW) could be also supplemented by re-injection of produced water, depending on the outcome of studies addressing temperature and solids-loading effects.

Another major benefit in using LSSW is the prospect of reducing the potential for hydrogen sulphide formation on breakthrough of injected seawater. By denying sulphate reducing bacteria a source of nutrients their activity can be limited.

16 ACKNOWLEDGEMENTS

The authors would like to express their thanks to Kerr McGee North Sea (UK) Limited management, Axsia Limited and the Janice licence partners for permission to present this work. Participants in the Janice Field Licence are Phillips Petroleum Company UK Limited, AGIP UK Limited, and Svenska Petroleum Exploration (UK) Limited.

References

1. Sinclair, R. and Weston, R.: *"Application of sulphate removal technology in Heron cluster field"*, paper presented at the Norwegian Society of Chartered Engineers Symposium, *"Offshore separation"*, 11 - 13 November, 1996, Kristiansand, Norway
2. Weston, R.: *"Nanofiltration: the use of membrane technology in offshore oil production"*, paper presented at International Desalination Association World Congress on Desalination and water re-use, 6 - 9 October, 1997, Madrid, Spain.
3. Hardy, J.A.: *"Control of Scaling in the South Brae Field"*, paper presented at the IBC Ltd. Conference, *"Advances in Solving Oilfield Scaling Problems"*, 7 - 8 October, 1992, Aberdeen, U.K.

SASOL's EXPERIENCE IN THE DESALINATION AND RE-USE OF ACID MINE DRAINAGE AND ASH WATER

J.G. Nieuwenhuis, G.H. Du Plessis,
M.P. Augustyn

B. Steytler,

Sasol Technology (Pty) Limited
Private Bag X 1034,
Secunda, 2302,
Republic of South Africa

Innovative Water Solutions (Pty) Limited,
P.O. Box 1210,
Secunda, 2302
Republic of South Africa

A.J. Viljoen,

I.W. Van Der Merwe,

Keyplan (Pty) Limited,
P.O. Box 430,
Kelvin, 2054
Republic of South Africa

Envig (Pty) Limited,
P.O. Box 7240,
Noorder Paarl, 7623
Republic of South Africa

1. INTRODUCTION

Since start-up of the Sasol Secunda Operations, a net accumulation of water was experienced in the ash water system (transport of ash from factories to the fine ash dams) and from mining operations (seepage of water through broken strata).

In line with Sasol Secunda's Water Policy where Sasol acknowledges its responsibility to conserve and protect all the water resources within the sphere of influence of its activities, membrane processes were installed to address water balance problems. Tubular reverse osmosis (TRO) followed by spiral reverse osmosis (SRO) was used to recover ash water and convert it to boiler feed water. Electrodialysis reversal (EDR) followed by SRO were installed to convert mine drainage water to boiler feed water.

These actions resulted in solving wastewater balance problems and simultaneously reduced raw water intake.

2. DESIGN

2.1 TRO/SRO Design

2.1.1 Pre-treatment section. A block flow diagram of the TRO/SRO plant is shown in figure 1. As pre-treatment, sulphuric acid is added to the ash water to control the pH between 3,0 and 6,5. This is followed by sand filtration to reduce the suspended solids concentration to *ca* 40 mg/ℓ. A scale inhibitor is added and the ash water is heated to 27°C. In order to limit biological fouling, chlorination is done twice a week during which a free chlorine level of *ca* 0,5 mg/ℓ is maintained for thirty minutes.

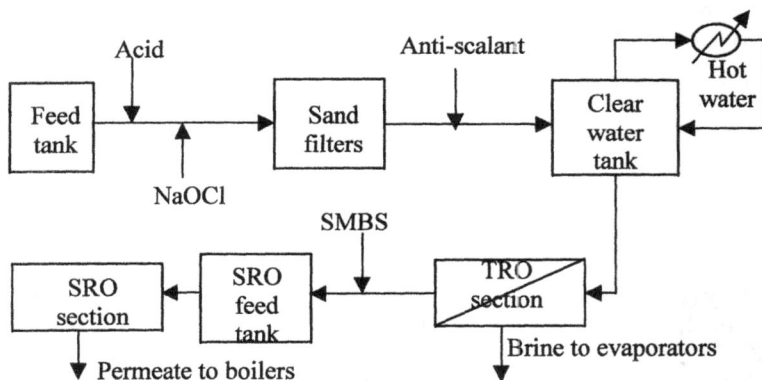

Figure 1 *Block flow diagram of the TRO plant*

2.1.2 Membrane section. The TRO section consists of eleven units. Each unit consists of 80 parallel branches each with 10 modules in series, giving a total of 8 800 modules. The plant is also equipped with a flow reversal mechanism and a sponge ball is shuttled through the membranes at thirty-minute intervals for in line cleaning.

The plant is operated at a constant water recovery of *ca* 40 %. The production capacity of each unit is 23 m³/h of permeate and the total capacity of the TRO plant is 250 m³/h. The waste stream is treated by means of three evaporators employing falling film and mechanical vapour recompression technology.

After TRO treatment the permeate is upgraded in two spiral reverse osmosis (SRO) units. Each unit has a design production capacity of 128 m³/h and water recovery is controlled at *ca* 90%. Standard high rejection, thin film composite polyamide membranes in a 10:5:3 configuration is used. Pre-treatment consists only of sodium-meta-bisulphite (SMBS) dosing to protect the polyamide membranes whenever chlorination is done at the TRO plant.

2.2 EDR/SRO Design

2.2.1 Pre-treatment section. A block flow diagram of the EDR/SRO plant is shown in figure 2. Pre-treatment in this case consists of clarification, sand filtration, cartridge filtration, anti scalant dosing, pH control using hydrochloric acid and heating by steam.

In the clarification step, $FeC\ell_3$ as well as cationic and anionic coagulation/flocculation agents are added to ensure proper clarification. As part of the pre-treatment step, potassium permanganate and sodium hypochlorite are added to oxidise manganese and iron to the insoluble oxide form.

Clarification is followed by gravity sand filtration and cartridge filtration (10 μm) to ensure that the silt density index is less than 5. The backwash from the sand filters is collected in a backwash recovery tank from which it is blended into the feed to the clarifier.

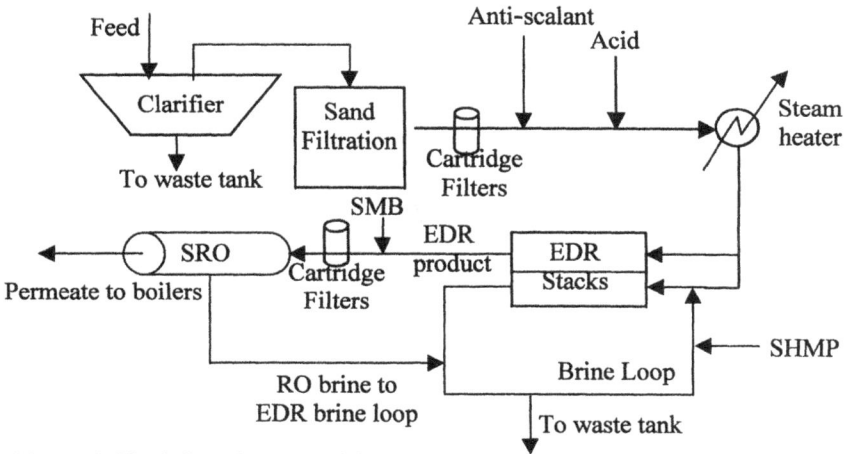

Figure 2 *Block flow diagram of the EDR plant*

Hydrochloric acid is added to reduce the pH to below 6,0. This is done to de-alkalise the water as much as possible before concentration in the desalination units. Temperature is controlled at 35°C to ensure optimum production.

2.2.2 Membrane section. The EDR/SRO plant is designed to operate at an overall water recovery of about 70%. The EDR section consists of two parallel trains i.e. one train with seven and the other with six parallel lines, each with three EDR stacks in series. The design production capacity is *ca* 250 m³/h. Standard EDR cation (Model C51) and anion (Model SXZL 286) membranes and the tortuous path spacers (Model Mk lll-3) are used in the EDR stacks. To ensure long electrode life, platinum plated titanium electrodes are used.

The plant is normally operated at a constant CaSO₄ saturation level of *ca* 250% in the brine circulation loop. This is done by a continuous purge from the brine loop. Make-up water to the brine loop is firstly supplied from the SRO brine and the balance is made up with pre-treated mine water. Sodium-hexa-meta-phosphate (SHMP) scale inhibitor is added to the EDR brine circulation loop to inhibit calcium sulphate scaling at the high saturation levels. The polarity is reversed every 20 minutes also to control scaling and fouling of the membranes.

The purge from the brine loop is rejected to the waste tank together with clarifier underflow to form the eventual waste from the plant.

After EDR treatment *ca* 80% of the salts have been removed. The permeate is then upgraded in two SRO units. Each unit has a production capacity of 125 m³/h and water recovery is controlled at *ca* 80%. Standard high rejection, thin film composite polyamide membranes in a 3:2:1 configuration is used. Pre-treatment to SRO is limited to cartridge filtration (5 μm). To protect the polyamide membranes, sodium-meta-bisulphite (SMBS) is added to remove any free chlorine. The brine stream is returned to the EDR as part of the make-up to the EDR brine circulation loop.

3   PLANT PERFORMANCE

3.1 TRO/SRO Plant

3.1.1 Pre-treatment section. Key performance criteria for the pre-treatment of TRO are pH, suspended solids concentration, feed temperature and planktonic and sessile biocounts. Results are given in table 1.

Table 1 *Pre-treatment performance*

Criteria	Target	Average	Maximum
pH	5.5	5.5 ± 0,3	5,8
SS mg/ℓ	50	75 ± 27	108
Temperature °C	27	25 ± 1,8	28,8
Planktonic counts cfu/ mℓ	Log 3	Log 4	Log 6
Sessile biocounts cfu/ cm^2	Log 2	Log 3	Log 5

It is evident from the results that pre-treatment has not always conformed to requirements. Problems were experienced with the performance of sand filtration and hypochlorite dosing.

Large variations in feed suspended solids (SS) concentrations were observed since start-up. The average feed SS concentration to the sand filters was 146 ± 33 mg/ℓ. On average, only 50% of the solids were removed, resulting in SS concentrations to the reverse osmosis section of approximately 75 mg/ℓ. Performance of the sand filters was further jeopardised by hydraulic shocks during the backwash cycle, resulting in filter nozzle damage and subsequent sand breakthrough [1]. The problems were however solved by replacing the existing filter nozzles with more suitable nozzles whilst the hydraulic shocks were eliminated by changes to the control system.

Initially sodium hypochlorite dosing was insufficient due to the degradation of the hypochlorite. This was overcome by reducing the inventory of sodium hypochlorite on site. Most recent results however again indicated elevated bioactivity. This was attributed to variations in feed quality and again degradation of the hypochlorite. Cost of sodium hypochlorite and SMBS (dosed at SRO to prevent oxidation of the polyamide membranes) merited a change to an inorganic biocide treatment program for the containment of biofouling.

3.1.2 TRO section. Key performance criteria for TRO treatment are permeate quality, salt rejection, standard membrane flux (flux at 4000 kPa and 25°C) and cleaning-in-place (CIP) frequency.

Considerable variation in feed water quality (Table 2) was observed since December 1995. Components such as calcium, barium, sodium, chloride and organic compounds (TOC) were significantly higher than the original design values. Despite this, TRO permeate quality was relatively constant and well within the specifications for SRO treatment.

Table 2 *Reverse osmosis feed and permeate quality*

Component	Feed concentration (mg/ℓ)	Permeate concentration (mg/ℓ)
TDS	3994 ± 786	96 ± 38
Calcium	422 ± 94	4,6 ± 3,2
Barium	0,2 ± 0.09	< 0,2
Sodium	917 ± 79	48 ± 7
Chloride	828 ± 238	44 ± 4
Sulphate	3254 ± 842	7.5 ± 5,1
Fluoride	18 ± 4.9	2 ± 0,3
TOC	52 ± 14	<10

Standard flux was on average 524 ± 65.5 ℓ/m^2.day, the variations (Figure 3) being indicative of changes in feed water quality. Flux was fairly stable especially considering the high fouling nature of the feed water and no permanent membrane fouling was observed since start-up. Using proper cleaning regimes (see below), it was possible to restore membrane flux to design values whenever it was needed.

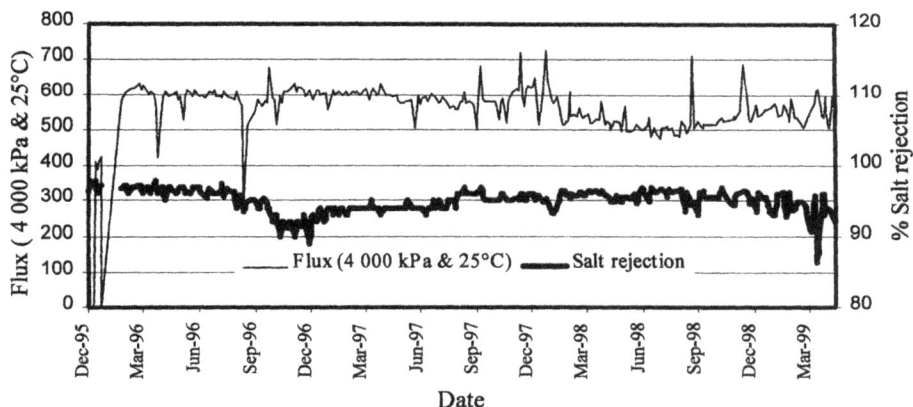

Figure 3 *Standard membrane flux and salt rejection*

On average, salt rejection of dissolved species as measured by conductivity, is 94,5% (Figure 3). This is slightly above the design value of 93%. A decline in salt rejection was observed after approximately eight months of operation from the initial level of 97% in December 1995 to as low as 90% in May 1997. This decrease concurred with inefficient cleaning of membranes and nozzle failure on the sand filtration section. After replacing the physically damaged modules resulting from nozzle failure, salt rejection was restored to 94%. However, the downward trend in rejection persisted. This was attributed to ineffective membrane cleaning. After implementation of a more effective CIP regime, salt rejection was restored to *ca* 94%.

Since February 1999, an increase in the barium feed concentration from *ca* 0,018 to 0,2 mg/ℓ (Table 2) was observed. This resulted in scaling on the membranes and subsequent membrane damage, which is evident from the decrease in salt rejection (Figure 3). Additional dosing of scale inhibitor solved this problem.

The feed water contains 52 ± 14 mg/ℓ TOC. Scheduled CIP's were thus primarily aimed at organic fouling and detergent cleaning was performed at thirteen-day intervals. During the first eight months of operation, CIP was not always efficient. Investigation proved that fouling was not only of an organic nature but a combination of organic and inorganic fouling. A study demonstrated that by alternating sodium-hexa-meta-phosphate (SHMP) and detergent CIP's, cleaning could be done effectively.

The high TOC concentration in the ash water creates the ideal conditions for biofouling. The sessile (attached) population is monitored weekly and is on average 10^3 cfu/cm^2. The dominant genera of the sessile population were found to be *Pseudomonas, Bacillus and Aeromonas*. No standard has as yet been established for the sessile population and can only be established when biofouling occurs. It was found that CIP with SHMP also contributed to the containment of sessile population and decreases of as much as 10^2 cfu/cm^2 were observed after CIP.

Planktonic (free-living) microorganisms are monitored weekly. Experience showed that a standard of less than 10^3 cfu/ml could be accepted for planktonic population. No permanent biofouling has as yet been observed which indicates the efficiency of disinfection.

3.1.3 SRO Section. Generally the plant performance was acceptable (Table 3). Flux was however slightly lower than design due to biofouling which could not be removed with standard CIP's. A CIP was developed (soak at high pH followed by flushing with a combination of compressed air and water) which successfully removed the biofouling, although at a higher CIP frequency. Since the introduction of an inorganic biocide dosing program, biofouling has decreased significantly. Biofouling is however a constant threat and the development of a treatment program and early warning systems are receiving attention.

Table 3 *SRO Performance*

Criteria	Target	Average
Water recovery (%)	90	$88 \pm 8{,}8$
Product Conductivity (μS/cm)	<30	$26 \pm 6{,}1$
CIP's / Train/month	1	2,4
Flux (ℓ/m^2.h)	25	$23{,}5 + 1{,}7$
Feed Pressure (kPa)	1 350	$1\,390 \pm 159$
Planktonic counts cfu/mℓ	Log 3	Log 5

3.2 EDR/SRO Plant

3.2.1 Pre-treatment section. Key performance criteria are Silt Density Index (SDI), iron and manganese concentrations. Monthly averages (Table 4) indicate that the plant performed well.

The exception is manganese removal that was on occasion not sufficient. This was attributed to a decrease in the mine water pH from *ca* 7,9 to 5,8 that resulted in poor oxidation of the manganese. Increasing the mine water pH to *ca* 7.5 by dosing soda ash resolved this problem.

Table 4 *Pre-treatment performance*

Criteria	Target	Average	Maximum
Fe – In (mg/ℓ)		0,37	0,62
Fe – Out (mg/ℓ)	0,2	0,1	0,18
Mn – In (mg/ℓ)		0,42	0,81
Mn – Out (mg/ℓ)	0,1	0,03	0,53
SDI – Out	< 5	3,9	>5

3.2.2 EDR Section. Key performance criteria here are water recovery, the number of stack cleans, salt rejection (percentage conductivity reduction) and product water quality.

From the results (Tables 5 and 6) it is evident that EDR treatment performed well. The objective of operating at a calcium sulphate saturation level of 250% was mostly met. Excursions did however occur from time to time. The plant however, operates with acceptable levels of manual stack cleaning. The target of six stacks per month, which represents cleaning each stack twice per year, is generally met or slightly exceeded. The plant was cleaned in place on a 3 – 4 week interval, which is also acceptable, considering the highly scaling operation conditions.

Table 5 *EDR Performance*

Criteria	Target	Average
Recovery	76%	79 %
Salt Rejection	78 – 82 %	76 %
CaSO$_4$ Saturation		
Max.	300	450
Average	250	262
Manual Stack Cleans	6	7

Salt rejection was slightly lower than the target but this was sacrificed in favour of higher water recovery. Despite this and the variations in feed water quality, which were within the specified limits, (Table 6), the product water quality was relatively constant and well within the design specifications.

Table 6 *Mine water feed, EDR and SRO product qualities*

Component	Feed concentration mg/ℓ	EDR Product mg/ℓ
TDS	3994 ± 786	1435 ± 438
Calcium	422 ± 94	36 ± 15
Sodium	917 ± 79	358 ± 151
Chloride	828 ± 238	121 ± 42
Sulphate	3254 ± 842	701 ± 487
TOC	2,12 ± 1,1	1.98 ± 0,4

3.2.3 SRO Section. Key performance criteria (Table 7) show that plant performance was acceptable. Flux was however lower than design due to biofouling which could not be removed with standard CIP's. A CIP using EDTA (1%), tri-sodium-polyphosphate (1%) and sodium-dodecyl-sulphate (0,5%) at 35°C and pH 10,5 proved to be very efficient. The target flux of 25 ℓ/m^2.h was difficult to achieve due to this unanticipated fouling. The plant was found to run very well in the region of 18 ℓ/m^2.h where long runs were achieved between CIP's. Biofouling is still considered as a serious threat and the development of a treatment program and early warning systems are being investigated.

Table 7 *SRO Performance*

Criteria	Target	Average
Water recovery (%)	85	79 ± 1,6
Permeate conductivity (μS/cm)	80	33 ± 9
CIP's / train/month	1	2,5
Flux - ℓ/m^2.h	25	20,1 ± 3,2
Feed pressure (kPa)	1 350	1350
Planktonic counts cfu/mℓ	Log 3	Log 6

The lower flux of 18 ℓ/m^2.h results in a somewhat lower recovery. This is however not considered being a problem, since the brine is recycled to the EDR plant as brine make-up and does not affect the overall plant recovery of 76 %.

4   OPERATING COST

Average operating costs since start-up were approximately the same for both combinations of processes at ca R3.50/m^3 of final product.

5   CONCLUSION

Up to now Sasol's experience with the treatment of wastewater using membrane technology, was positive. Plants conformed to expectations and the original objectives were met. Where problems did occur, they could be resolved.

In the Sasol water systems there is more potential to apply membrane technology and this will most probably realise in the future.

References

1   J.G Nieuwenhuis, B. Steytler, I.W. van der Merwe, M.P. Augustyn,
        The Water Institute of Southern Africa, Biennial Conference, Cape Town, 1998,
        Vol 3, Section 3E-1.

RECOVERY OF WOOL SCOURING EFFLUENT UTILISING MEMBRANE BIOREACTOR (MBR) TECHNOLOGY AS PART OF THE ACTIVATED SLUDGE SYSTEM FOLLOWED BY TWO–STAGE REVERSE OSMOSIS (RO) MEMBRANE CONCENTRATION

A. R. Bennett

Membrane Systems Division
ACWa Services Limited
ACWa House
Keighley Road
Skipton BD23 2UE
United Kingdom

1. INTRODUCTION

The first fully automated commercial facility in the United Kingdom to incorporate a membrane bio-reactor (MBR) enables up to 85% of water resources required on site to be recovered for use in the wool scouring process by using high pressure reverse osmosis (RO) membrane technology.

The site utilises approximately 13 m^3h^{-1} of clean water into the wool scouring process on a continuous basis. A zero effluent discharge consent has driven investment in the effluent treatment and recycling system with all wastes being concentrated and tankered away for disposal.

The system incorporates the Zenon ZeeWeed™ microfiltration (MF) membrane system as part of the activated sludge process. Koch - Fluid Systems high pressure RO membranes are included to concentrate filtered water from the MBR and recover 85% as permeate with the remaining 15% waste containing concentrated contaminants. The 15% loss is made up from the local mains water supply.

1.1 Development of MBR Technology

Early work on MBR technology in the 1960's by Dorr-Oliver in the USA was not commercialised – it was the forerunner of Zenon Municipal Systems that developed a commercial system in the 1970's containing tubular ultrafilters based on a single sludge aerobic – anoxic process (Husain & Côté, 1999). System sizes were small treating flows generally below 100 m^3day^{-1} with plants used for treating and recycling wastewater for small companies.

It has only been in the last ten years that MBR technology has moved from the small-scale applications to flows in excess of 10,000 m^3day^{-1}. The pace of development has increased more rapidly in the last five years, along with the increasing number of applications for MBR technology (Churchouse & Wildgoose, 1999). The number of plants and flows treated by the Kubota process provides a good example of the rapid surge in demand from the marketplace. In 1993 four Kubota plants had been installed, the largest treating a flow of 125 m³day⁻¹. This had risen to 150 with a maximum capacity of 1907 m³day⁻¹ in 1998 and the total to date for 1999 in June was 237. The largest plant was now up in the mega litres per day (Mld⁻¹) category at 12,700 m³day⁻¹ (12.7 Mld⁻¹).

Three reasons have been sited by Churchouse and Wildgoose for this expansion:-

1. Design fluxes have increased.
2. Membrane life has increased from 3 to 8 years.
3. Increased levels of manufacture have reduced fabrication costs.

The result is that MBR technology is now able to compete with traditional effluent treatment technologies, and the MBR is now available to solve various effluent problems.

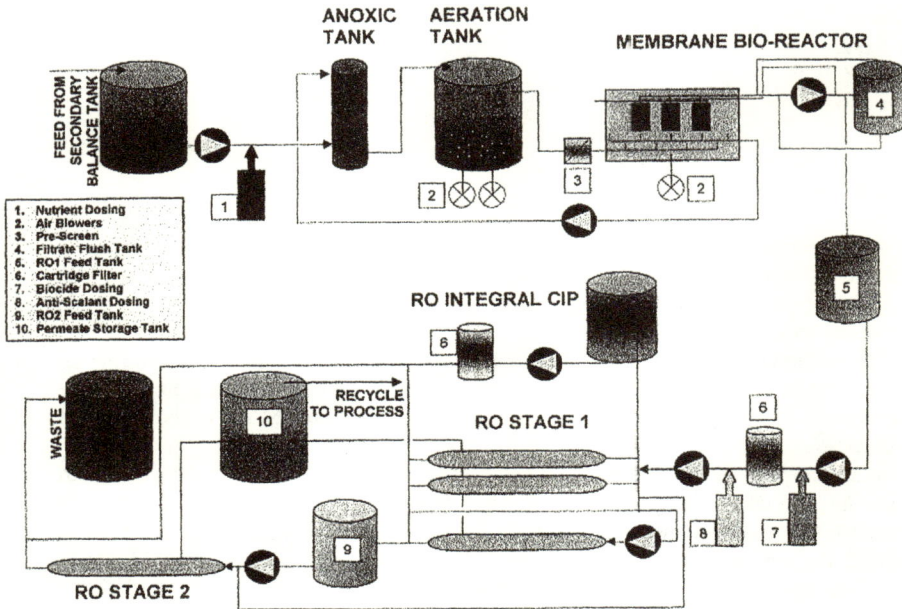

Figure 1   *Effluent Treatment System Schematic*

1.2 RO Developments

The RO industry has been steadily expanding and there are now a number of major players offering various types of RO membrane elements for applications ranging from desalination to wastewater reuse. Use of RO in effluent treatment has been restricted in the United Kingdom but a substantial amount of work has been undertaken in the United States and the Far East utilising RO technology in conjuction with microfiltration as an effective pre-treatment.

Recent developments leading to higher surface area elements has reduced the capital cost of installations and improving manufacturing techniques have lead to a reduction in the feed pressures required resulting in lower power consumption and the consequent reduction in operating costs. In parallel to these positive developments has been the improvement in permeate (product) quality achievable due to increased rejection of dissolved ions. Work with RO in the wastewater field has shown this membrane technology to be a consistently reliable technique for separating an effluent stream into a purified permeate for re-use and a concentrated membrane reject for disposal.

1.3 Combination of Technology

This project has involved the combination of the MBR and RO technology in a manner that has capitalised on the recent developments of the membrane separation processes described above. It has only been recently that both technologies have been economically attractive enough to be included in a capital plant project, and this application is the first commercial facility to be installed in the United Kingdom.

2. PROCESS DESCRIPTION

The process installed in depicted in Figure 1 on the previous page. It comprises activated sludge with an anoxic zone and an MBR with product from this further processed by the RO system.

2.1 Traditional Activated Sludge Enhanced by MBR Technology

Supernatant from the dissolved air flotation (DAF) process is pumped to the aeration tank of the activated sludge system via an anoxic tank for nitrification. The typical analysis of the feed to the plant supplied by ACWa Services Limited is shown in Table 1 below. The chemical oxygen demand (COD) entering the process following wool scouring and DAF reaches 5000 mgl^{-1} with a total dissolved solids (TDS) of up to 8000 mgl^{-1}. Return activated sludge is mixed with the DAF feed in this anoxic tank and the facility exists to add chemical nutrients as required to promote biological growth in the activated sludge system. Following oxidation of the biomass the mixed liquor gravitates to the MBR.

This application of the ZeeWeed^{TM} MBR process enables the separate anoxic zone to be accommodated outside of MBR reactor vessel. Also, the separate bio-reactor enables the development of an active mixed liquor with levels of suspended solids up to 12,000 mg/l as opposed to a typical 4,000 mgl^{-1} found in traditional activated sludge systems. The separate MBR unit also enables easier cleaning-in-place (CIP) of the membranes. Husain & Côté (1999) have also described the MBR system without the additional bio-reactor capacity. This configuration is mainly used for straightforward municipal wastewater applications.

The micro-filtration membrane system is rated to remove particles nominally above 0.1 μm in size and the filtered product is drawn up under vacuum through 24 membrane elements packed with polypropylene based hollow fibres. These elements can be seen suspended in the reactor vessel in Figure 2. The separated biomass is pumped back to the anoxic tank as return activated sludge and periodically waste activated sludge is removed from the system to maintain the correct sludge age.

Table 1   *Influent Design Basis*

Parameter	*Value / Unit*
Volume	312 m^{3}day^{-1}
Flowrate	13 m^{3}h^{-1}
Chemical Oxygen Demand (COD) Concentration	3,500~5,000 mgl^{-1}
Biochemical Oxygen Demand (BOD) Concentration	2,000~2,500 mgl^{-1}
Suspended Solids	< 300 mgl^{-1}
Settled Solids	< 200 mgl^{-1}
Total Dissolved Solids	< 8,000 mgl^{-1}
Conductivity	< 20,000 μscm^{-1}
Ammonia	200~400 mgl^{-1}
Total Nitrogen	< 600 mgl^{-1}
Total Phosphate	< 0.2~0.4 mgl^{-1}
Iron	< 80 mgl^{-1}
pH	7.0~7.5
Temperature	30 to 35 °C

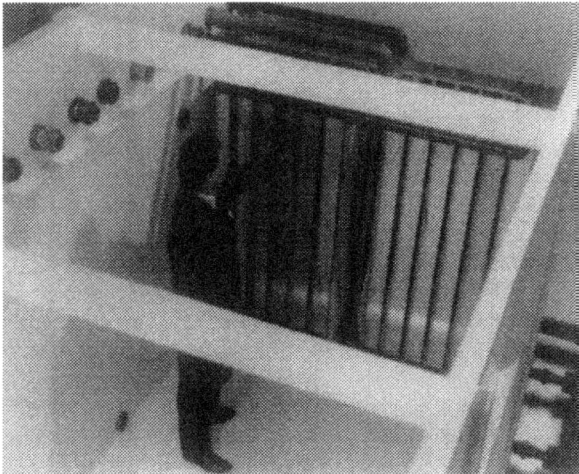

Figure 2   *ZeeWeed^{TM} Microfiltration Unit*

2.2 RO Two Stage Concentration

Filtered water from the MBR enters the first stage of the RO plant. The feed is pressurised to 40 bar g by an inverter controlled multi-stage centrifugal pump to maintain a fixed feed flowrate before entering a 2:1 array of pressure tubes each containing six high rejection thin film composite membranes supplied by Koch – Fluid Systems. The pressure is boosted between the two arrays by a fixed speed centrifugal pump.

The RO process works by pressurising the feed side of the membrane with a pressure greater than the natural osmotic pressure of the solution. Where the natural process of osmosis results in the movement of water from a high osmotic pressure to a lower osmotic pressure through the semi-permeable membrane, the application of a net driving pressure in excess of the natural osmotic pressure reverses the process to produce a small volume of concentrated solution. The amount of permeate produced divided by the volume treated is termed the recovery of the RO system. A two stream RO plant provided for treatment of a borehole supply is depicted in Figure 3 on page 225.

The first stage recovers 70% of the MBR filtrate as permeate for re-use and utilisation in the wool scouring process. The typical permeate quality required is shown in Table 2 below. An Allen-Bradley programmable logic controller automatically controls the recovery rate of the RO plant along with all other control interlocks for safe and efficient operation of the system. Operator monitoring and control initiation is provided via a user interface screen.

The concentrate stream from the first stage RO is further processed by the second stage RO plant. A fixed feed flowrate is maintained using a positive displacement plunger type pump capable of generating pressures up to 80 bar g. The feed is thus pressurised onto a single pressure tube comprising five membranes suitable for the saline concentrate from the first stage of the RO system. 50% of this concentrate is recovered from the second stage to add to the first stage permeate. The overall recovery of both RO systems in this two stage concentrator is thus 85%.

RO permeate recycled to the wool scouring process is recovered at a continuous rate of 11.05 m^3h^{-1} with a TDS of less than 300 mg/l and a COD of less than 10 mgl^{-1}. The level of dissolved solids is less in the recycled water than in the mains supply.

Table 2   *RO Permeate Typical Specification*

Parameter	Value / Unit
Feedwater flowrate	$13m^3h^{-1}$
Permeate flowrate	$11.05\ m^3h^{-1}$
Total hardness	$<25\ mgl^{-1}$
Silica	$<0.5\text{~}3.0\ mgl^{-1}$
Iron	$<0.1\ mgl^{-1}$
Manganese	$<0.02\ mgl^{-1}$
Alkalinity	$<50\ mgl^{-1}$
Total Dissolved Solids	$<300\ mgl^{-1}$
Colour	<20 Hazen

3. OPERATING REQUIREMENTS

At the time of writing this document the plant has not been operating for a sufficiently long period to be able to provide detailed operating costs for the process. However, it is hoped to provide details at the Conference in March.

3.1 Nutrient Dosing

The provision for nutrient dosing into the feed from the DAF plant has been provided but this has not been utilised to date. It would be possible to add chemicals providing a combination of N P and K to enhance the biological activity in the activated sludge plant and the MBR.

3.2 MBR Cleaning

The MBR control logic allows for backflushing of the hollow fibre membranes on a periodic basis generally set at one hour intervals. Hypochlorite can be introduced into the backflush water to provide a chemically enhanced backwash and this is generally undertaken on a weekly basis.

The trans-membrane pressure (TMP) of the system is typically expected to increase from 0.4 to 1.2 bar g over a period of two months following which dilute backflushing with acid and / or caustic solutions will be required.

3.3 RO Dosing and Cleaning

An anti-scalant is continually dosed at 2.5 mgl^{-1} concentration in the feed to the RO plant in order to minimise scaling on the concentrate side of the membrane and also to minimise biological fouling.

In addition to the anti-scalant, every week for 30 minutes a non-oxidising biocide is dosed at 400 mgl^{-1} concentration. This shock dose prevents the formation of biofilms on the RO membrane surface and has the effect of increasing the period required between cleaning-in-place (CIP).

CIP chemicals will be required on a monthly basis and the facility exists to use either high or low pH proprietary solutions for periodic recirculation through and soaking of the RO membranes. A solution is made up manually in the CIP tank and heated up to 40 °C. A small centrifugal pump is then used to force the CIP liquid through the plant. Effective CIP procedures will reduce the net driving pressure required down close to initial requirements thus prolonging the membrane life.

4. CONCLUSIONS

The integration of the MBR and RO membrane technologies alongside the traditional activated sludge process has resulted in an efficient system allowing the recovery of water for re-use on-site. This has provided the advantages of reduced mains water consumption, a 6-fold reduction in waste disposal costs in order to comply with the zero discharge consent and the recycling of RO permeate with a lower dissolved solids content than the mains water supply.

Figure 3 *Two Stream Reverse Osmosis System built by ACWa Services Limited*

References

1. H. Husain & P. Côté, *The Zenon Experience with Membrane Bio-Reactors for Municipal Wastewater Treatment*, Published in MBR-2 Conference Proceedings, Cranfield University, 1999.
2. S. Churchouse & D. Wildgoose, *Membrane Bio-Reactors Hit the Big Time:- from Lab to Full Scale Application*, Published in MBR-2 Conference Proceedings, Cranfield University, 1999.

PERFORMANCE ON A REAL INDUSTRIAL EFFLUENT USING A ZenoGem® MBR

D. Mallon and F. Steen

Anglian Water Services Ltd.,
Thorpe Wood House,
Peterborough,
PE3 6WT

K. Brindle,

School of Water Sciences,
Cranfield University,
Cranfield,
Beds,
MK43 0AL.

Abstract

A Zenon ZenoGem® pilot MBR (membrane bioreactor) was studied using submerged ZeeWeed® MF (microfiltration) membranes in activated sludge, treating effluent from a chicken processing factory. The performance of the MBR was monitored with regard to the final permeate quality, pre-treatment requirements and potential re-use opportunities. The paper assesses the performance and compares the advantages and disadvantages of the process together with likely costs. The influent had a COD between 2,000 - 4,500 mg/l, BOD average 1853 mg/l, with high SS (suspended solids) and high ammonia content. The COD, BOD, ammonia and phosphate removal efficiencies were higher than 92%, 99%, 95% and 98% respectively. Phosphate removal/assimilation was greater than expected; this was thought to be due to the acclimatisation of the biomass to high phosphate levels, around 50 mg/l, in the chicken blood in the influent. The effluent was of a high enough quality to discharge directly to river making this process competitive in price with other conventional processes. The effluent was unsuitable for reuse within the factory.

Keywords

Membrane bioreactor; wastewater treatment; activated sludge, chicken processing effluent.

1 INTRODUCTION

In conventional aerobic activated sludge systems, biomass separation from the treated liquor relies on sedimentation of aggregated mixed microbial flocs. If biomass separation from the treated effluent is facilitated by physical retention within the bioreactor the need for flocculation is removed. The necessary operating conditions can be achieved with cross-flow membrane bioreactors where the membrane acts as a filter to provide better than clarified effluent. Progress in the engineering of membrane bioreactors and the development of immersed membrane configurations, has produced compact, robust, cost-effective systems. Immersed membrane systems use either hollow fibre membranes or flat sheet membranes (Bussion et al. 1998)[1] and consist of membrane modules submerged directly into the activated sludge compartment. Typically they operate in an outside-inside filtration mode at low suction pressures, (1-5 bar). Operation at low suction pressure means a reduction in the transmembrane pressure. This means reduced fouling of the membrane and slower formation of a cake layer, resulting in less frequent backwashing and lower operating costs, as smaller pumps are required. However, because the force driving the liquid through the membranes is relatively low, a large membrane surface is required to compensate for the low flux rates achieved. Sludge wastage is minimised by maintaining a low F/M ratio while the footprint of the plant is reduced by operating at high biomass concentrations, typically 15 - 20 g/l (Cote and Pound 1997)[2]. In addition, biomass retention allows build-up of a waste-specific microbial population, of particular use when dealing with industrial effluents, thereby providing the most effective biological treatment. The main improvements arising from the coupling of membrane technology with biological wastewater treatment are summarised in Table 1.

Table 1. *Conventional and membrane biological treatment comparisons.*

Problems and limitations associated with conventional biological treatment	Improvement from the inclusion of membranes in the process
Variations in the pollutant loading	Acceptance of variation in concentration of activated sludge including high concentrations
Very slow rate of kinetic reaction	Natural selection and total retention of the bacterial population
Settlement of bacteria in clarifier can rate limit process	Removal of bacteria from effluent not dependent upon settling characteristics of bacteria
Post-treatment removal of viruses required prior to reuse	Significant removal of viruses from effluent make direct reuse possible
Insufficient contact time between macromolecules and micro-organisms	Increase in contact time allowing effective treatment of low biodegradable products
Rate of sludge production creates a problem for disposal	Sludge volume reduced with minimised cost of post-treatment
Irregular quality of treated water and absence of effective barrier to bacteria	Production of high quality effluent, free from bacteria, offering the possibility of recycling
Bulky	Foot-print reduced

1.1 Materials and Methods

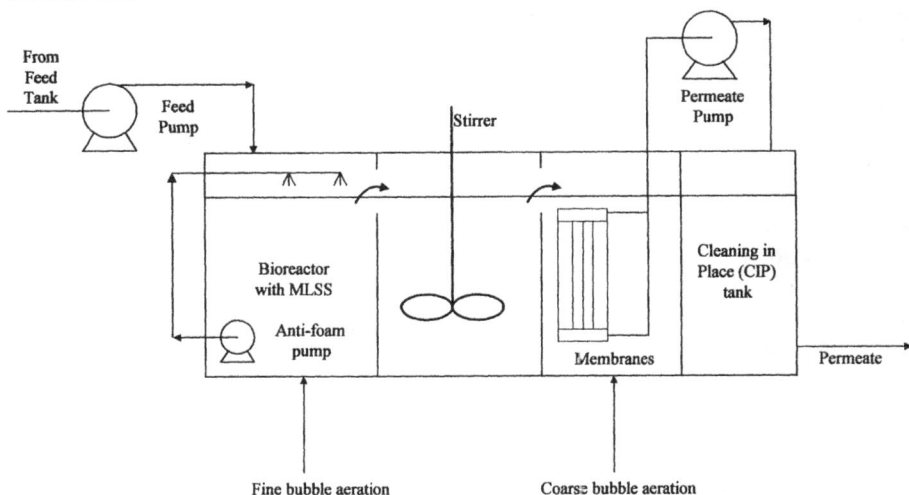

Figure 1. *Schematic diagram of the pilot membrane bioreactor.*

All chemical analysis was carried out on site with a Dr Lange colorimeter using standard methods. These results and all microbiological analysis were verified at a NAMAS accredited laboratory using standard methodologies[3,4,5]. Bacteriological samples were monitored for *E.Coli* and Coliforms and viral samples were monitored for Human Enteric Viruses.

A ZenoGem® pilot plant was fed with effluent from a chicken processing factory that was high in COD, BOD, ammonia, suspended solids and fat. It also contained blood and feathers from the culling process. The effluent was screened through a 2 mm drum and then through two crude fat removal tanks, (Figure 3). These consisted of 1m³ tanks operated with a constant overflow, the feed being taken from the bottom of the tank using a small submersible pump. The pilot plant used in this study had a total working volume of 4 m³, (Figures 1 and 2). It consisted of two reactors in series (of 3 m³ and 0.8 m³, respectively), followed by a membrane compartment, and CIP (cleaning in place) tank, operated under aerobic conditions. Fine bubble aeration was supplied at the bottom of the first bioreactor and the second was mechanically stirred. Two hollow fibre membrane modules, (surface area 13.8m³ each, pore size 0.1 microns supplied by Zenon GmbH), were submersed in the third compartment. The membranes are chlorine resistant made from plastic, woven to obtain the required pore size. Hollow fibre membranes provide a higher surface area to volume ratio than flat plate membranes, thus occupying less reactor volume. Coarse bubble aerators were positioned at the bottom of the membranes supplying air to increase cross flow velocity and increase mass transfer in the vicinity of the membranes. The membranes were backwashed with permeate for 40 seconds after every 300 seconds of operation. This is to prevent surface fouling and dislodge the cake layer formed at the membrane surface. Provision was made in the pilot plant for chemical cleaning with sodium hypochlorite, however a chemical clean was not required during this study. A biomass recirculation pump was in constant operation transferring biomass from the membrane compartment back into the first reactor. The pH was maintained between 6.75 and 7.25 by acid and alkali dosing.

Figure 2. *Picture of the pilot plant*

Figure 3. *Picture of first grease trap*

1.2 Feed Quality

The characteristics of the screened feed are given in Table 2.

Table 2. Feed characteristics.

	Average	Minimum	Maximum
Total COD (mgO$_2$/l)	3181	1280	5064
Total BOD (mgO$_2$/l)	1652	156	2950
Ammonia (mgN/l)	20.78	0.58	84.8
Total Phosphate (mgP/l)	19.20	1.87	62.9

2 RESULTS AND DISCUSSION

2.1 Plant Operation

The bioreactor was seeded from a local sewage works; the biomass concentration slowly increased over the first month. Initially foaming inhibited the development of a significant microbial population. This was overcome by drip-feed application of anti-foam oil and a reduction in aeration rate. However, overnight foaming caused the MBR to shutdown when influent detergent levels rose due to factory cleaning. This was reduced by using a timer switch on the first fat removal tank pump to ensure feeding only during production hours. Foaming stopped when the MLSS reached approx. 10 g/l and sludge levels increased rapidly, around 1g/l/d, to a maximum of 28 g/l. The suction pressure delivered by the pump dropped from 1 bar to 0.68 bar at MLSS of approx. 15-18 g/l, indicating a rise in transmembrane pressure (Figure 4). The increase in MLSS was accompanied by a decrease in DO from over 2.5 mg/l to below 0.4 mg/l and a deterioration of permeate quality with the COD rising from 100 - 200 mg/l to over 350 mg/l. Maintaining the DO is critical to maintaining permeate quality; unfortunately, increasing the air flowrate through the blowers on the pilot plant would have produced large rather than fine bubble aeration and a subsequent loss in oxygen transfer. This is not expected to be a problem on a full size plant, which would incorporate DO control.

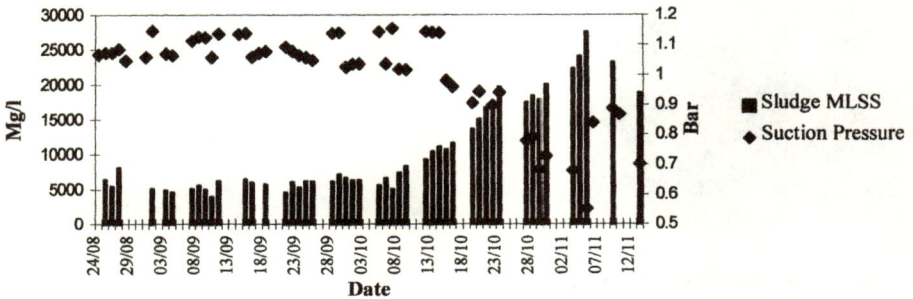

Figure 4. *MLSS concentration and Suction Pressures.*

Permeate quality was of a consistently high enough standard to discharge directly to river, BOD levels typically less than 10 mg/l, with an average 92% and 99% removal of COD and BOD, respectively, but had a slight colour from the blood in the influent, (Figure 5). The lower influent COD values around the 22nd October correspond to the timer on the feed stream being removed and samples being taken around midday, when the plant was producing permeate from a less concentrate overnight feed. The lower COD values on the 9th and 10th of November correspond to low influent COD concentrations. Flow balancing should ensure a more consistent and better quality permeate. Samples of the permeate showed bacteria and virus concentrations were below the level of detection.

Ammonia and phosphate removal efficiencies were excellent, with an average of over 95% and 98% removal respectively, (Figure 6). Average ammonia in the permeate was 0.86 mg/l and average phosphate was 0.23 mg/l. The phosphate removal was particularly good and showed a marked decrease from the start of October onwards, which was accompanied by a slight rise in ammonia concentration in the permeate. The cause is not fully understood but may be due to one or more of the following;

2.2 Results

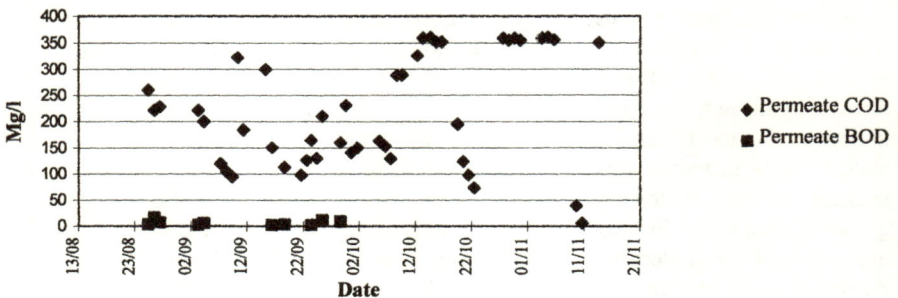

Figure 5. *Permeate COD and BOD concentrations.*

1. Variation of the feed concentrations of ammonia and phosphate. Again reinforcing the need for buffering upstream of a full size plant.
2. Biological phosphate assimilation.
3. Simultaneous nitrification and denitrification caused by an anaerobic nucleus forming in the MLSS, which at higher concentrations became more viscous.
4. Concentrated blood containing up to 50 mg/l phosphate being added at the start of the study, conditioning the MLSS microbial population enhancing phosphate removal when this high concentration was later removed.

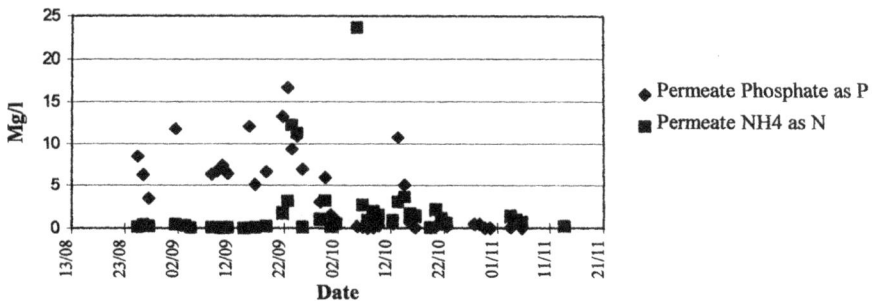

Figure 6. *Permeate ammonia and phosphate concentrations.*

Loading rate on the bioreactor was 6.2 kg $COD/m^3/d$, 3.2 kg $BOD/m^3/d$ and 0.9 kg $COD/m^2/d$, 0.5 kg $BOD/m^2/d$ on the membrane. Samples taken from the permeate and mains water supply to the site showed a large increase in the concentration of dissolved salts in the permeate. Chlorides rose from 68 to 172 mg/l Cl and sodium rose from 47.6 to 127 mg/l Na. Reuse of the permeate from the MBR in the factory, without further treatment, would create an environment in the bioreactor that would be toxic to the microbial population. However, the permeate was treated using a Reverse Osmosis (RO) plant and caused no fouling to the RO membranes.

2.3 Cost Benefit Analysis

The processing plant where the MBR was trialed currently has DAF pre-treatment before discharge to sewer. Trade effluent bills for a flow of approx. 900 m^3/d are circa £300 K per annum and volume discharged is expected to double. Effluent treatment plant on site requires a small footprint, as space is at a premium, and low odours, as the factory is opposite a residential area.

Estimated costs of a ZenoGem® membrane bioreactor to treat 1800 m^3/d are:

Capital Expenditure - £1,400,000   Operating Expenditure -18.3p/m^3 permeate
The permeate from the pilot MBR was of a high enough quality to discharge direct to river and savings in trade effluent costs give an estimated payback period of approx. 3 years.

3 CONCLUSION

This study monitored the performance of a Zenon ZenoGem® with a real industrial effluent. Problems occurred with foaming of the biomass that stopped when then the MLSS reached 10 g/l. Suction pressures dropped at operation over 15 g/l. The permeate was of a high quality, COD and BOD 242 mg/l and 6 mg/l, respectively, with no visible suspended solids. Ammonia removal efficiency was 95.8% with an average concentration in the permeate of 0.86 mg/l. Phosphate removal efficiency was excellent and greater than expected at 95.9%, with an average concentration in the permeate of 0.23 mg/l. The permeate was unsuitable for reuse in the factory without further treatment. However, preliminary costings show that a direct discharge to river would make an MBR plant a viable economic option for treating this case study waste.

4 ACKNOWLEDGEMENT

The authors are grateful to Frank Duty and Marilena Demetriadi at Anglian Water Services Ltd. for their help in taking samples and assistance with maintaining and running the pilot plant.

References

1. H. Buisson, P. Cote, M. Praderie and H. Paillard. The Use of Immersed Membranes for Upgrading Wastewater Treatment Plants *Wat. Sci. Tech.* Vol. 37, No. 9, pp. 89-95, 1998.
2. P. Cote and C. Pound Results of Pilot Studies for the Reuse of Municipal Wastewater Presented at the Workshop on Membranes in Drinking Water Production - Technical Innovations and Health Aspects, L'Aquila, Italy, June 1-4, 1997.
3. HMSO 1994 Report 71: Microbiological Analysis of Drinking Water.
4. HMSO Methods for the Isolation of Human Enteric Viruses from Waters and Associated Materials.
5. HMSO publications Methods for the Examination of Waters and Associated Materials (Blue Books).

MEMBRANE TECHNOLOGY IN WOOD, PULP AND PAPER INDUSTRIES

Jørgen Wagner

Osmonics Desal
DK-2820 Gentofte,
Denmark

1   INTRODUCTION

This document will describe three membrane applications used in the wood, pulp and paper industries:

- Reverse Osmosis (RO) of Medium Density Fibreboard (MDF) effluent using spiral wound membrane elements.
- Ultrafiltration (UF) and RO of Spent Sulphite Liquor (SSL) using a combination of a PCI tubular UF system and a Desal™ high temperature spiral wound membrane element RO system.
- UF and Nanofiltration (NF) of white water using a combination of the plate and frame CR filter from Valmet Flootek and Desal spiral wound NF elements designed for high temperatures.

The pulp and paper industry, much like others, is characterised by extremely large process flows and high volumes of fresh water consumption. Membrane filtration offers the possibility to reduce water consumption, reduce solids discharge or simply enhance existing equipment such as an evaporator.

1.1   RO of MDF Effluent

This process is one of the success stories of recent years. What started as a dream was realised in the creation of a zero discharge facility from an otherwise polluting plant.

MDF is a wood based panel, similar to particleboard but much more versatile since it can be machined (planed, turned and routed). It is generally manufactured from softwoods, such as pine. MDF can be produced according to a wet and to a dry method. According to the wet method wood is first chipped and washed to remove any extraneous debris such as sand. The chips are then steamed to soften the fibres and then fed into a refiner. Excess liquid (MDF effluent) is separated from the fibres, which are then mixed with glue and flash dried. The dried mixture is then distributed onto a forming line to create a uniform mat and then finally a heated continuous press presses it to around 1/40th. The final product can then be used with or without a coating such as laminate. Applications include furniture, kitchens, laminate flooring and house internals such as skirting boards and picture rails.

The MDF effluent is traditionally treated in a biological treatment plant. However, a very high Biological Oxygen Demand (BOD) load and formaldehyde from the glue kitchen makes it difficult to treat and results in poor treated effluent quality. A biological plant occupies much space and consumes a lot of energy in the aeration system. Consequently, there are several incentives to investigate alternative processes that operate more economically. RO membrane technology was identified as such a process.

Esmil Process Systems LTD, a long-time Osmonics partner, started test work in 1993. The objective was to develop a process that recovers all suspended and dissolved solids from the MDF effluent, and allows reuse of the solids recovered. Both goals were achieved.

Pre-treatment was, and continues to be, a challenge. It is very difficult for a spiral wound membrane element to handle the fibres that inevitably are present in the MDF effluent. It was also realised that there were various naturally occurring resins in the water which could disturb the membrane process, and which are difficult or impossible to remove by conventional cleaning methods.

Esmil chose flocculation for pre-treatment that proved extremely challenging. Extensive test work was required to identify a flocculent which achieved the pre-treatment goals without damaging the membranes which are typically susceptible to the cationic nature of effective flocculent. Esmil was successful in identifying a satisfactory flocculent although precise dosing became a necessity. If dosing is too low excess suspended solids will be fed into the membrane system. If dosing is too high polymer may blind the membranes and membrane flux will suffer.

After flocculation, the liquid undergoes several filtration steps. First, it is sent through a large filter press, the filter cake from which is incinerated to produce energy for steam production. The filtrate from the filter press is fed through a sand filter followed by cartridge filters which act as protective, safety filters more than anything else since physical blocking of the spiral wound elements could easily occur if suspended solids were to accidentally by-pass. Although this extra pre-treatment process may sound expensive, it is, in the author's opinion, absolutely essential to ensure that the RO plant is not damaged by debris that may enter as a result of upstream failures. Although they are rare, failures eventually occur, and it is better to be safe than sorry. The sand filter is back-flushed regularly, but the flush water is not discahrged to sewer as usual - it is sent back to the MDF raw effluent tank for flocculation and RO treated again.

The filtered MDF liquid is now treated by an RO system. All dissolved solids are concentrated. The concentrate is simply sprayed on the fibres before pressing. Two advantages are achieved by doing this. First, it is an excellent disposal route (there is no effluent to treat). Second, and more surprising, is that it actually improves board quality and saves a little glue since the sugars polymerise during heating and create a glue-like material.

The RO plant is cleaned regularly. The spent Cleaning In Place (CIP) liquid is sent back to the MDF collection tank, which is against all the rules of membrane filtration, but in this case it works!

The RO permeate is the last stream to discuss. It typically contains <100 mg/L of COD, meaning it can be used in a number of factory applications thereby reducing the consumption of fresh water.

The operating parameters are not extreme considering pressures do not exceed 40 bar and temperatures average 30°C.

In summary, the MDF effluent and all the chemicals and water used in the plant may be divided into the following three components:

Table 1 *Typical MDF data*

	Feed	Permeate	Concentrate
Volume	6,7 m³/h	<0,7 m³/h	>5,6 m³/h
°C	around 30 °C		
SS	20,000 mg/l	n/d	
COD	20,000 mg/l	<150 mg/l	200,000 mg/l

1. Filter cake, which is incinerated
2. RO concentrate, which is used as glue
3. RO permeate, which is reused in the factory.

1.1.1. No water or solids are wasted. The plant design employs the traditional multistage recirculation design (Figure 1). Although it has been argued that the single-pass design used for water treatment is an alternative; it is, in the author's opinion, not a viable alternative since flux must be stable over time in order for single-pass designs to succeed. The flux in this application is quite variable, which would make the operation of a single-pass system extremely difficult.

Pumps are, as usual, also a challenge. The feed pump(s) is the easiest to specify since it simply supplies a specified flow at a given pressure. Grundfos pumps were chosen and have operated fine. Although the mechanical seals have been problematic, they have actually performed better than expected considering the pump was manufactured to pump cold, clean water only.

Figure 1 *Membrane Skid*   **Figure 2** *Plant Overview*

The recirculation pumps, on the other hand, have been a problem. It is tempting to use borehole pumps, such as the BM type from Grundfos, since they are inexpensive. But in the long run the problems exceed the savings and they eventually were abandoned. Problems were as follows: First, the system operating temperature was on the high side for the pump, although within specifications. Since the liquid it pumps cools the pump, it is likely that the warm water caused the motor to overheat. Second, the seal between the pump motor and the product leaked occasionally, allowing MDF liquid to enter the motor which is obviously undesirable. Lastly, the cables were not sufficiently insulated, resulting in short circuits and motor burnout. More conventional pumps, albeit with quite expensive mechanical seals, have proven to operate better.

An indication of this application's success is the fact that Esmil has installed a total of eight plants in Europe since the first plant was installed in Chirk (see Figure 1 and 2), and the number is still growing.

1.2   UF and RO of SSL

Pulp production by cooking with highly acidic Ca-sulphite is the first and oldest method used industrially. Today, there are very few of these mills because the sulphite process has largely been replaced by alkaline cooking, otherwise called Kraft pulping. Sulphite pulping survives because cellulose made from the sulphite process produces a specific type of cellulose, which is difficult or impossible to produce by other processes.

For every ton of pulp produced, one ton of solids are dissolved. This means that a factory producing 300.000 TPY of pulp also generates 300.000 TPY of waste water containing approximately 55% lignosulphonate (LS), 25% various sugars and 20% ash by products. The sheer tonnage can flood most markets since there are few buyers that can absorb so much product. The problem of selling huge amounts of a byproduct and the difficulty of finding a market for materials such as lignosulphonate has resulted in the limited use of membrane technology. Borregaard (Norway) is the exception because they employ membrane filtration plants world wide and dominate the lignosulphonate market. There are other markets, but they have for non-technical reasons not been exploited yet.

Nevertheless, there are a few RO systems treating Spent Sulphite Liquor that can be reported upon. RO of SSL can potentially replace evaporators, and can certainly enhance the performance of existing evaporators. For many reasons it is the latter which has been most successful, although few systems have been installed.

One particular application concentrates permeate from a UF plant treating SSL. An RO membrane system may either treat SSL or UF permeate, even though the composition of SSL varies considerably. Table 2 shows typical values for SSL and for UF permeate.

The UF plant treating SSL is a tubular system from PCI. The system incorporates 12 mm tubes, and the main advantage of the tubular system is its ability to handle suspended solids, especially fibres, and to do so with a minimum of pre-treatment. It is common to use a 20K MWCO PSO membrane in such an application. The operating temperature is around 70°C, and the majority of the LS are rejected by the UF membrane - only the very low MW LS may pass through the membrane. However, LS is not of great concern since LS is easy to wash away and does not influence flux very much.

Table 2. *Typical data for the RO plant*

	Feed to RO	RO concentrate	RO permeate
Total solids	6,9%	14%	<0,1%
Conductivity	7500 µS	11500 µS	600 µS
Osmotic pressure	8	16	

The RO system concentrates the UF permeate from 7% TS to 14% solids prior to evaporation. The TS in the feed can vary significantly, and the high operating temperature of 70°C is a challenge for the membrane elements. However, the use of Desal™ Duratherm® Excel membrane elements, and the specification of a 3,8" OD, has secured good mechanical stability for the elements. It should be noted that 3,8" is a commonly used diameter in the dairy industry, but it is certainly unusual in this kind of industry. This element was chosen for its excellent mechanical strength: it is probably the sturdiest element on the market. However, it is likely that a new plant built today would use 8" elements since the designs, and mechanical strength, have improved considerably over the last few years.

The plant is built as a two-stage recirculation plant. The plant employs 400 m² of membrane surface area, and the water removal capacity is a nominal 10,000 L/h that was in accordance to requested capacity. It turned out that the UF plant operated at considerably higher capacity than expected, which means the RO system was too small. In order to treat all the UF permeate as fast as it is produced, a significant enlargement is needed.

For many months the system performed as expected and did not show any tendency to lose flux or foul. Lately, however, the RO plant has experienced a peculiar problem of low capacity. Since UF permeate is the feed, the pre-treatment is almost perfect, and no harmful chemicals from the SSL should be reaching the RO membranes. If a UF membrane failed, raw SSL could penetrate, but it is rare to incur a failure so massive that it seriously affects the membranes.

The present problem can probably be traced to the water supply since the plant uses river water that contains humic acid. Humic acid is colloidal in nature, and there may be clusters housing microorganisms, which secrete fatty or sticky products as part of their metabolism. When such water is heated, such as for steam production, the fatty part is separated from the humic acid, and the result is water, which contains a material behaving much like mineral oil. Some of these undefined oily substances have been observed, and it has had a rather devastating effect on polyamide membranes. Since pH during cleaning is limited to 11,5, it is difficult or impossible to remove the fouling layer. We are still investigating how or if the membranes can be regenerated, and how to prevent such problems in the future.

1.3 UF and NF of White Water in the Paper Industry

White water is a term used to describe the wastewater produced during the formation of paper in the wet end of a paper machine. The water is rather white due to the high content of suspended solids.

Paper machines have grown to almost incomprehensible size and the following are a few figures that are literally breathtaking. New machines form paper measuring up to 10 meters in width. The speed of the paper is approaching 2000 meters per minute, which is equivalent to 120 km/h, or 75 mph. The largest machines produce 40 TPH paper. Now take into account that the pulp slurry contains less than 2% cellulose fibres, meaning that the volume of white water is around 1800 m³/h. The price for a machine is well over 200 million US dollars.

These numbers reveal two important aspects concerning this application. First, the paper industry's wastewater problem is serious and very large. Second, an operator of such a machine will not install any equipment that endangers the operation of production: downtime is extremely expensive.

In order to achieve the proper paper quality, it is necessary to add several chemicals to the pulp slurry, and some of those in quantity. For instance, it is common to add $CaCO_3$

as filler to obtain the glossy paper used in magazines. The pulp slurry is poured onto a fast moving mesh where the formation of paper takes place. One of the major challenges in making paper is keeping the paperforming portion of the machine in good working order. Water is drained first by gravity, and then by vacuum, leaving the pulp. After a few seconds of dewatering, the paper is lifted off and continues into the paper machine. The wire is then subjected to intense cleaning by nozzles, and is then ready again for paper formation.

Cellulose fragments some of the filler and some of the chemicals can become wedged in the wire mesh and must be removed. Otherwise, the wire will lose its ability to drain water from the paper, and machine capacity will be limited. The quantity of water needed for spraying is staggering, and it is a challenge to clean the white water by conventional methods to such an extent that it can be used in the nozzles. For years, UF has been used to do just this, and has performed well. However, low flux and low availability, combined with too little paper machine knowledge, has prevented the widespread adoption of UF.

In order to demonstrate that UF is a viable, practical process for the filtration of white water, a full-scale plant must be built since a small-scale pilot will prove nothing. Valmet Flootek (former ABB Flootek and Raisio Flootek) has, for more than 5 years, been operating full-scale UF systems on white water. It has been proven conclusively that UF meets all the criteria for use in the harsh environment. It should be noted that Valmet is one of the world's major producers of paper machines. Therefore, they can combine their paper know-how with the membrane know-how of Flootek to effectively sell a complete and proven system to the end user. This combination of knowledge is unique and necessary for the success of this application.

The filter used by Flootek is called the CR filter. It is a plate and frame type, with approximately 84 m^2 per filter. The CR filter gains sufficient turbulence by mechanical means, and does not rely on pumps. The filter employs something similar to pump impellers between each plate. The advantages include:

- The volume to be treated and the linear velocity over the membrane are independent variables
- The investment is well under US $2 per m^3 water treated
- The volumetric concentration ratio can be varied, and it can be very large without affecting the operation
- The module is not sensitive to suspended solids (within reason)
- High, stable flux over extended periods of time is possible.
- The disadvantages include complex module construction.

The volumetric concentration ratio is approximately 25 meaning that 100 litres of feed are divided into 96 litres of permeate and 4 litres of concentrate. It is known that the permeate recovery can be increased to >99%, however, the extra volume of water recovered is quite small. It is most probably not worth while to attempt higher levels of recovery since the solids content in the concentrate increases enough that it impairs the function of the CR filter.

The author has stated in the "Membrane Filtration Handbook" that flux is never >100 Lmh for any length of time. The CR filter has broken that rule by maintaining flux of 200 - 300 Lmh for weeks on end. This is, of course, not valid for all products, but is for most. A major reason for this occurrence is the high shear rate on the membrane surface.

There are several full-scale systems in operation. It has been proven conclusively that the UF permeate is very well suited for nozzle water and for reuse in other parts of the paper process. The CR filter has achieved an important goal in papermaking: reduction of water consumption.

1.3.1 NF of white water UF permeate The more water that is recycled in a pulp or paper mill, the more one has to watch out for salt build-up. One major step forward in the pulp industry is replacement of Cl_2 bleaching by oxygen bleaching because this replacement removed a major source of chloride. Chloride is feared in all pulp and paper operations because chloride can so easily cause corrosion, that at best results in poor paper quality and at worst, can stop a complete system.

Many newer mills have recycled their water to the point that they need a method to remove salt from the system. Nanofiltration and Reverse Osmosis can achieve this objective. For several years, Flootek has been piloting NF of UF permeate and the results have always been good. However, the first priority was to gain acceptance of UF. Now that that has taken place, the next step is to gain acceptance of NF and/or RO.

As this paper is being written in 1999, a full-scale system is being commissioned. The data presented below reflects the results from the pilot test. Data from the full-scale, on-line system will be available during the conference in March of 2000. Tables 3 and 4 show data obtained from the full-scale UF systems and the pilot-scale NF system.

Table 3 *Typical values. Kirkniemi. Papermachine 1 and 2*

		pH	µS	COD_{Cr}	TS	Salts	m³/h	Bar
UF	Feed	5.1	482	3282	7317	4460	188	<4
	Concentrate	5,0	489	6347	10978	6754	180	
	Permeate	5,2	492	1210	4319	2823	8	
NF	Feed	5,0	492	1210	4913	2823		<15
	Concentrate	4,9	1216	4209	14495	10743		
	Permeate	5,0	202	305	1246	345		
UF	Reduction factor		0	2,71	1,69	1,58		
NF	Reduction factor		2,44	3,97	3,47	8,17		

Table 4 *Typical Concentration of Solutes*

	Ca	Mg	Al	Fe	Cl	SO_4	SiO_2	meq/l	mS
Feed	175	8,9	2,0	3,3	44	1046	60	0,18	230
Concentrate.	700	38	5,0	13	112	3658	106	0,53	649
Permeate	1	0,1	0	0,1	23	0,3	56	0	32
Reduction	99%	99%	100%	97%	48%	100%	7%	100%	86%

Some comments regarding the system are as follow:
- The NF flux varies considerably. It appears that an average flux of 30 Lmh is reasonable at 50°C and 12 bar. These operating conditions are gentle, and the probability of mechanical membrane damage is quite small. It is most likely that flux stability can be improved by increasing the cross flow. This is one of the parameters that will be investigated on the full-scale system.
- Cleaning every 3 to 4 days is recommended in order to prevent the flux from

decreasing too much.
- The concentration ratio is close to 4, meaning that 75% of the feed becomes permeate while the remaining 25% are discharged as concentrate.

The NF shall enable the factory to recycle much more water and therefore consume less fresh water. The bigger the NF system, the more water can be recycled. The full-scale plant will be used to investigate the relationship between water recycling and the impact on the water composition inside the paper machine. UF will make it possible to reduce fresh water consumption from $10m^3$ per ton pulp to $5m^3$. NF will make it possible to reduce fresh water consumption to $2m^3$ per ton pulp possible that is equal to zero discharge. Both are major achievements in water conservation.

2 FINAL REMARKS

Membrane technology is finally being used in the pulp and paper industry almost 30 years after the first very optimistic market studies predicted that membranes would be used on many different pulping liquids. UF of SSL and UF of bleaching effluent (E1) were the only processes that utilised membrane technology for two decades. It is interesting that 20 years passed before Esmil and Flootek realised new processes, while the known processes largely lay dormant. The author is confident that there are many more possibilities for membrane technology in this industry. However, it appears to require an unusual combination of people, membrane technology, pulp and paper companies, and, most importantly, lots of hard persistent work to achieve success.

References

1   Finnemore, S. "Achieving Zero Output" and Private Conversations, Esmil Process Systems Ltd, 1998.
2   Tepler, M., et al. PM and BM White Water Treatment with Membrane Technology. Raisio Flootek OY, 1998.
3   Tepler, Milan. Private Conversations, Valmet Flootek, 1997, 1998, 1999.
4   Wagner, J. Membrane Filtration Handbook. Wagner Publishing, Copenhagen, 1998.

CASE STUDIES OF WASTEWATER RE-USE FOR THE PETROCHEMICAL, POWER AND PAPER INDUSTRY

Bruce Durham

USF Memcor
Vivendi Water
Wirksworth DE4 4BG, UK

1 INTRODUCTION

A revolution is taking place in the water industry. The environmental issues combined with ever more stringent health expectations and the Ofwat directives are driving these changes as our industry moves towards common carriage and a better deal for the customer. How closely will we follow the example of the Gas and Electricity industries into an open and highly competitive market? Water companies are already benchmarking their utility and asset management business against the Gas and Electricity suppliers.

Reuse of wastewater is actively being encouraged by this water revolution. But we need innovators to champion these beneficial solutions, agreement on who owns the water, logical financial incentives to interest the budget holders and reuse quality guidelines to protect all concerned. The UK has been slow to take advantage of new opportunity due to the priority set following privatisation and lack of funding in a highly regulated water industry.

It is easy to understand why California, Australia, Spain, South Africa and Israel are ahead of the UK. It is not so easy to understand why the Netherlands, Belgium, France, Scandinavia and Germany are also more advanced in reuse projects than the UK. This is especially difficult to comprehend when East Anglia is an arid region and we even have water shortage problems in high rainfall areas in the UK due to limited catchment reservoirs and no opportunity to build more reservoirs.

Best practice studies need to consider international track record not just the limited amount of work that has been completed in the UK.

This paper is intended to introduce and demonstrate the viability of large reuse projects in the UK based on long term international operation experience on reuse projects for petrochemical, power and the paper industry.

2 REUSE TECHNOLOGY

Reverse osmosis is the key process used in the reuse of wastewater to remove the dissolved salts. RO membranes are designed to operate under steady-state conditions.

Sources, such as wastewaters are susceptible to rapid changes in suspended or dissolved solids loadings and are complex waters to treat.

An RO element is not designed to handle suspended solids, as the RO membrane retains the solids. This results in rapidly declining flow and shortens the membrane life.

Continuous microfiltration (CMF) systems are being used in over 40 installations to protect RO systems. There are two key elements to the CMF process - the 0.2μm pore size hollow fibre membranes (out-to-in filtration) and patented air backwash system, both of which offer a number of benefits:

♦ Very high feed solids can be accommodated with no effect on the system performance other than an automatic change in the frequency of the backwash.

♦ A consistent filtrate flowrate and quality is continuously sustained, with an automatic process, irrespective of the quality of the feed water (Silt Density Index (SDI)<3, Turbidity<0.1 NTU)

♦ Energy consumption is very low - typically 0.15-0.3 kWh/m^3 of treated water. Even with several hundred mg/l of feed suspended solids, cross-flow is not required.

♦ The compressed air backwashing is much more effective than the conventional liquid backwashing at removing contaminants due to high energy and good distribution. It also offers the added benefits of discouraging grow-through and avoids the need for prechlorination of the backwash liquor, which is normally a prerequisite of conventional liquid backwashing membrane systems.

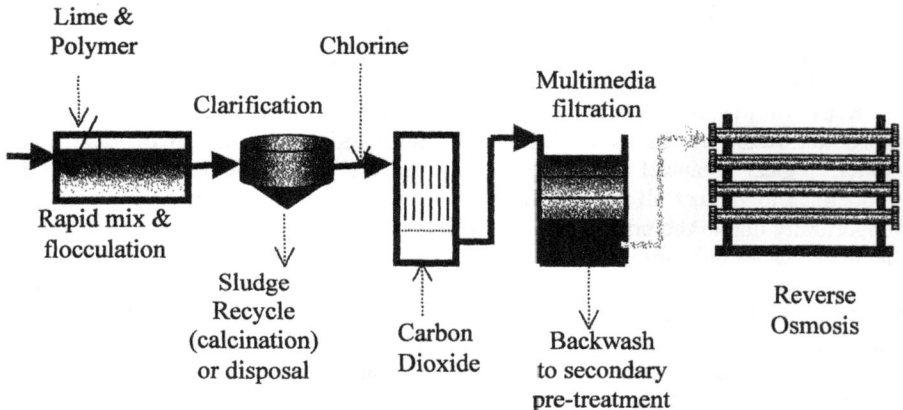

Figure 1 *"Conventional" RO Pretreatment*

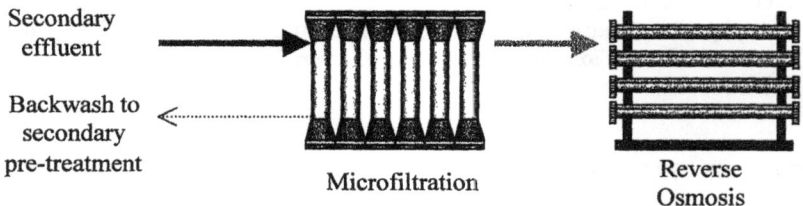

Figure 2 *Continuous Microfiltration RO Pretreatment*

The following sections review three case studies.

3 WEST BASIN WATER DISTRICT, CALIFORNIA

The confidence in CMF/RO technology developed through the work done at Water Factory 21, [1] led to the two West Basin wastewater reuse projects.

Over-abstraction has depleted the aquifer, causing saline ingress. In addition, industrial activity has caused further pollution making the aquifer unusable. There is complete dependence on expensive imported water.

Secondary sewage from the Hyperion works is now used in a variety of ways to alleviate this. In one application, commissioned in 1997, Memcor processes are used in 15 Mlitre/day CMF/RO system to supply the nearby Mobil and Chevron refineries with feed for their boiler water plants. In an earlier project Memcor supplied an 11.5 Mlitre/day CMF plant as pretreatment for RO to provide water used for injection in a barrier scheme to keep out further sea water.

Figure 3   *CMF & RO system at West Basin*

CMF has simplified the wastewater reclamation process by eliminating the lime handling, addition and recovery systems and the flocculation, clarification, recarbonation and filtration processes, resulting in lower chemical requirements and eliminating the production and disposal of solid waste sludge.

3.1 Membrane Life

The frequency and cost of replacing the membranes is of key importance to the viability of a project, especially when membrane cost is typically 15% to 25% of capital cost.

The CMF system at West Basin, which contains 5 units, automatically checks the integrity of 9 million membrane fibres as part of the operating procedure. Two sub-modules have been repaired in the last two years of operation.

The following capital and operating costs comparison for the conventional and microfiltration pretreatment to RO were presented in March 1999 at the AWWA membrane Conference. [2]

The operation costs for the two pretreatment systems were calculated from July 1997 to October 1998.

Table 1   *Cost Summary*

Description	Conventional Pretreatment		Microfiltration	
	(a) £	(b) £/ML	(c) £	(d) £/ML
Fixed Costs				
Capital Costs	586,851	84	276,720	78
O & M Labour	197,446	28	49,362	14
Replacement Parts & Supplies	21,875	3	49,269	14
Subtotal Fixed Costs	806,172	115	375,351	106
Variable Costs				
Chemical Costs	490,581	54	62,637	17
Sludge Production & Handling	310,467	34	7,960	2
Power	122,916	13	49,994	13
Subtotal Variable Costs	923,964	101	120,591	32
Total Fixed & Variable Costs		216		138

Table 2   *Quality Summary*

	Conventional	**Microfiltration**
Feed Turbidity (NTU)	4.5 – 7.5	4.5 – 7.5
Filtrate Turbidity	0.12 – 0.32	0.06 – 0.10
Comparison on SDI	4.2 – 6.5	0.8 – 2.2

Table 3   *Process Design Scope*

	Conventional	**Microfiltration**
CAPITAL	Clarifier Trimedia Filtration Cartridge Filters RO	Strainers Microfiltration Cartridge Filters RO
VARIABLE	Lime Ferric Chloride Sodium Hypochlorite Sulphuric Acid Scale Inhibition	Sodium Hypochlorite Sulphuric Acid Scale Inhibition

4 ERARING POWER STATION, AUSTRALIA

This is a municipal/industrial re-use project in which secondary treated effluent from a local Sewage Treatment Plant undergoes tertiary treatment at Eraring Power Station in New South Wales. The plant design combines Memcor continuous microfiltration (CMF) and reverse osmosis (cellulose acetate membranes) to produce high purity water for all power station uses except for drinking and showering.

The power station is located on the shore of Lake Macquarie, a tidal saline lake near the coast and approximately one hour from Sydney. The project consisted of connecting the outlet from the Dora Creek sewage treatment works to a new membrane treatment system incorporating microfiltration and reverse osmosis. The treated sewage is then used instead of potable water for boiler feed, cooling water, dust suppression and fly ash handling.

4.1 Process Design

This 2-stage membrane system has been in operation since March 1995 producing $63m^3/h$ with a planned expansion to $168m^3/hr$. Effluent from Dora Creek Sewage Treatment plant flows under gravity to centrifugal pumps, which deliver the feedwater via a single in-line motorised 500 micron self cleaning strainer to two Memcor CMF units.

Figure 4   *CMF System*

Virtually all suspended solids, faecal coliforms and giardia cysts are removed. The SDI of the CMF filtrate water is consistently reduced to 1.5 allowing RO membranes to be operated at approximately 40% higher flux than is possible with a pretreatment based on the conventional lime coagulation/sedimentation/filtration process.

Filtrate from the CMF units is dosed with sodium hypochlorite, for downstream control of biological growth. Sulphuric acid is also dosed to reduce pH and minimise hydrolysis of the cellulose acetate RO membranes. Microfiltered water is drawn from the storage tank, dosed with antiscalant and passed through a 5 micron disposable cartridge guard filter for feed to the RO plant.

The RO system comprises two trains (2 x 50%), each with two stages - the first stage comprising 6 RO pressure vessels and the second stage comprising 3 (6:3 array). Each pressure vessel houses seven RO membrane elements. The membrane elements themselves are 8.5-in. diameter x 40 inches long cellulose acetate membranes rated at 98% salt rejection. The RO membrane, allowing only water to pass through, rejects salts and organics. Permeate (treated water) which is virtually free of all salts and micro-organisms, is piped to a degasser tower

Figure 5   *RO System*

to increase pH by flashing off CO_2 and is then stored in a $60m^3$ treated water tank.

The system provides the following benefits:

1)   The treated water (permeate) from the RO plant has a TDS of < 40 mg/l when treating secondary sewage of 500-1500 mg/l TDS . The existing deioniser capacity has been increased by more than 3 fold as a result of the membrane pretreatment resulting in over A$150,000 savings per annum in chemical costs.

2)   The membrane system removes bacteria and viruses for health risk minimisation and to protect the downstream equipment from biofouling.

3)   RO permeate can be used directly for washdown, dust suppression and gland seal, with no health risks.

4)   The compact size of the installation enabled an existing building to be used.

5)   The total system is fully automatic, operated through the SCADA system and has self diagnostic data logging.

6)   Membrane integrity tests automatically take place daily. Clean in Place (CIP) frequency is determined automatically by the control system. All cleaning chemicals are recovered and reused automatically on the microfiltration system.

7)   All effluents from the system are recycled or used for dust suppression.

5 ZERO DISCHARGE PAPER MILL, NEW MEXICO, USA

One of the best known "Zero Effluent" Paper mills is McKinley Paper Company's mill in New Mexico, USA, producing test liner from OCC (old corrugated container). U.S.F supplied the water reclamation plant project with Continuous Microfiltration (CMF), RO and with Crystallization (USF HPD S.A.).

The mill has operated since 1994 and it was originally designed for production of 135 000 ton/year of 100 % recycled fiber based board.Today the production is approximately 165 000 ton/year.The raw water consumption is approximately 1.5 m^3/ton of product. This raw water is reverse osmosis treated deep well water from the Plains Electric Generation and Transmission company next door. There is no foul sewer near the mill in order to discharge the effluent and the raw water availability is restricted. These matters were the original driving forces for the "Zero Effluent" concept.

The effluent from board production and OCC handling is treated biologically in an SBR (Sequenced Batch Reactor) type of activated sludge treatment plant. The biologically treated effluent is aerated and clarified in the same basin, in sequences. During clarification the aeration is simply turned off and the sludge is settled to the bottom of the basin. Excess sludge is removed from the bottom for dewatering and the clear effluent is removed from the surface of the basin. The water flow to the biological treatment is approximately 1500 m3/d. Typical COD content of the water is 3000-5500 mg/l in the inlet, TSS approx. 250 mg/l. COD in the outlet is 450-500 mg/l and TSS 40-50 mg/l. The effluent is cooled to suitable level (< 37 °C) before the SBR treatment in order to optimize the biological activity.

5.1 Membranes After Biological Treatment

The biologically treated effluent contains approximately 40-50 mg/l suspended solids of which is removed by the CMF microfiltration system.

The CMF treated water has TSS < 1 mg/l and SDI < 3, which is very suitable for reverse osmosis treatment, in order to remove salts. Reverse osmosis treatment in McKinley is for approx. 660 m3/d, with 75% recovery, resulting in approx. 1000 m3/d water recycled after microfiltration. The CMF treated water is low in BOD and COD, low in TSS (< 1mg/l), low in bacterial (10^3 CFU/ml).

An HPD Crystallizer evaporates the concentrate from the RO plant in order to turn the "last liquid form of effluent" to solid waste. All TDS left over from biological, CMF and RO treatment is concentrated to RO brine. The amount is approx. 168 m3/d with TDS concentration as high as 8000 mg/l.The Crystallizer is a tube type of falling film evaporator where the driving force for evaporation is generated by compressing the vapor pressure higher and simultaneously heightening the vapor temperature with mechanical single stage centrifugal vapor compressor. The condensate from the crystalliser is the cleanest treated water fraction from the whole water reclamation plant and it is re-circulated back to the process.

The final solid waste, practically 80 % Na_2SO_4 salt crystals, is disposed with the rest of the solid wastes by incineration in the multi fuel boiler in the coal mine next door.

6 CONCLUSIONS

1. Continuous Microfiltration allows Reverse Osmosis technology to treat previously impractical source waters, as microfiltration allows RO feedwater quality to be controlled and consistent.

2. CMF produces a constant high quality effluent that is independent of feed water variations.

3. Microfiltration allows the membrane inventory of an RO plant to be reduced by up to 30 to 40%.

4. Microfiltration membranes have simplified the RO pretreatment process resulting in lowering the operating costs of both the pretreatment and RO systems.

5. CMF is proven to be more economical on whole life costs and is helping to drive down the cost of ownership of RO installations.

6. CMF allows TFC RO membranes to operate in a stable condition on municipal wastewater.

References

[1] G.L Leslie et al. , 1997. Pilot testing of microfiltration and ultrafiltration upstream of reverse osmosis during reclamation of municipal wastewater.

[2] W.Won et al., 1999 Comparative life cycle costs for operation of full scale conventional pretreatment/RO and MF/RO system, AWWA Membrane Conference, Long Beach, March 1999.

[3] G. Craig, Overview of the long term application of wastewater reuse at a major powerstation. April 1999 .

[4] T Pohjolainen, 1999 Case study of 'Closed' board mill and application studies in Finland.

Posters

PRACTICAL EXPERIENCE WITH A MEMBRANE BIOREACTOR FOR WASTEWATER TREATMENT-SEMI-CROSS-FLOW ULTRAFILTRATION

S. Geißler, K. Vossenkaul, Th. Melin

Institut für Verfahrenstechnik,
der RWTH Aachen,
Turmstr. 46,
52056 Aachen
Germany

P. Ohle, E. Brands, M. Dohmann

Institut für Siedlungswasserwirtschaft,
der RWTH Aachen,
Mies-van-der-Rohe-Str1,
52056 Aachen
Germany

1 INTRODUCTION

Currently, the use of ultra and microfiltration for retention of activated sludge is receiving increasing attention. Using ultra- and microfiltration has the advantage of optimising the discharge hygiene. Further advantages are the low sludge generation and decreased tank volume. The challenges of using ultra- and microfiltration in the municipal wastewater treatment technology are to reduce investment and operating expenses and to cope with inflow variations. The development of new methods of operation is necessary to fulfil these requirements.

2 PILOT PLANT

The pilot plant concept consists of a biological stage with nitrification and denitrification as well as a membrane plant with two module units. The bigger unit consists of five cushion modules with approximately 40 m² in total. This module unit is supposed to guarantee the operation of the biological stage. One single module is running in parallel to the first unit to optimise the operating parameters of the membrane plant and membrane materials.
Using external modules it is possible to run the plant more flexibly than with submerged modules. The vertical installation of the membrane modules allows an air sparging from the bottom into the modules. This air is used to induce shear forces, hereby reducing cake formation to achieve high filtrate fluxes. To generate the necessary driving permeate sided partial vacuum is applied.

3 RESULTS

During the pilot tests the operational procedure of the plant is gradually adjusted to realistic conditions of local sewage plants, i. e. variations in inflow conditions, different solid content, etc. The variation of operating parameters like pressure, feed velocity and solid content allows different flux rates to meet capacity requirements.

The flux was varied between 10 and 45 l/m²h. Each set of operating parameters was realised at stable conditions for several weeks at low energy consumption. For short periods the flux can be raised up to 70 l/m²h. The energy consumption for a fluxrate of 30 l/m²h was 1 kWh/m³.

References

1. P. Ohle, E. Brands, K. Voßenkaul, S. Geißler" Optimierte Integration der Membrantechnik in die biologische Stufe kommunaler Kläranlagen"; 2. Aachener Tagung Siedlungswasserwirtschaft und Verfahrenstechnik 9/98

TREATING HIGHLY COLOURED WATERS:DESIGN INNOVATIONS AND IMPLICATIONS

A.B.F.Grose.   D.Welch.

Thames Water Research & Technology,   PCI Leopold Ltd.,
Spencer House   1 Kilduskland Road,
Manor Farm Road,   Ardrishaig.
Reading, RG2 OJN.   Argyll, PA30 8EH
UK   UK

E.Irvine.

West of Scotland Water,
Engineering Services,
John Macdonald House,
296 St. Vincent Street,
Glasgow. G2 5RU. UK.

The supply of drinking water to much of Scotland is characterised by numerous remote, small, rural supplies, many of which are affected by coloured water. The colour is imparted by the naturally occurring organic substances, humic and fulvic acids, which upon chlorination, result in formation of the disinfection by products (DBPs) known as trihalomethanes (THMs). The high domestic and wild animal populations mean that *Cryptosporidium* and high bacteriological counts are also endemic to many of the catchment areas.

The use of membranes to treat these high coloured waters in Scotland is now well established. To date there are 20 operational *Fyne* membrane plants all sized at less than 1 Ml/d. Although the *Fyne* tubular membranes have proven successful for small treatment plants, spiral wound membranes remain the most cost competitive option for larger installations.

If spiral wound membranes are to be adopted as a successful and cost effective option for larger treatment works, there are several major issues to be addressed:

- Spiral membranes easily become plugged by particulates in the feed source, this necessitates the incorporation of an efficient pre-filtration stage which significantly increases system costs.
- Spiral wound membranes require chemical cleaning at regular intervals- the disposal of this chemical waste can be both problematical and costly.
- Traditionally spiral wound membranes have operated at high pressure, however with the advent of low pressure membranes on to the market, energy costs can be significantly reduced.

Eight years ago a team Comprising of West of Scotland Water (WOSWA),

Leopold (a division of PCI Membranes Ltd) and Thames Water Research & Technology developed the tubular *Fyne* Membrane Process – now a market leader in Scotland. A purpose built 0.3 Ml/d pilot installation treating coloured water is now in place at WOSWA's Amlaird WTW.

In the graph below a comparison can be seen between the colour of the raw (feed) water to that of the final (product) water at the pilot plant at Amlaird WTW.

Raw colour v's Final colour

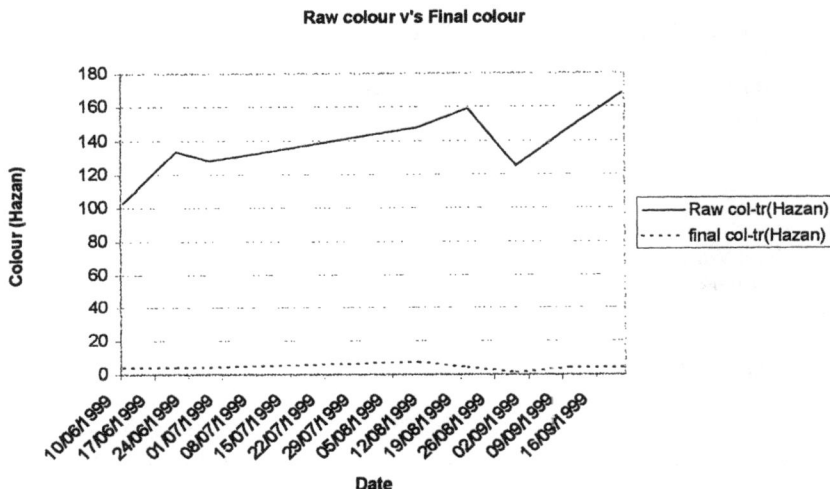

Figure 1 *Raw vs. Final colour*

Other determinands show similar levels of removal. The concentrations of Aluminium, Calcium, Iron, Manganese and THM's in the product water all comply the water quality legal limits.

However, to keep up with demands, there must be continued development of the technology. For example, the emergence of other high rate, compact, automatic and reliable pre-filtration systems are being evaluated against traditional pre-filtration systems. Also, different membranes are now available including a low pressure cellulose acetate (CA) membrane that has been granted DWI materials approval. The particular benefit of this type of membrane is its capacity to remove THM precursors.

Evaluation of membrane life continues allowing longer life guarantees thereby reducing cost of compliance.

TREATMENT OF LEACHATE BY THE MBR PROCESS (MEMBRANE BIOREACTOR)

Tony Robinson

Stork-Protech (UK) Ltd
Sheraton House
2 Rockingham Road
Uxbridge
Middlesex
UB8 2UB
UK

When considering the biological treatment of industrial waste water, thoughts are generally directed towards conventional activated sludge, biofilter, membrane techniques, etc. There is a new process, at least relatively new to the UK, using the MBR (Membrane BioReactor) to combine the advantages of aerobic treatment and membrane separation.

The MBR, fundamentally, consists of two integrated unit operations.

- A high concentration BioReactor with an MLSS in excess of 20,000 mg/l. The BioReactor can be pressurised thereby improving the solubility of oxygen in the biomass.
- An ultrafiltration plant where the membranes are used to separate biomass from final effluent.

The advantages of the MBR process can be summarised as follows:-

⇒ Excellent quality effluent with potential discharge to a water course.
⇒ Ability to handle difficult wastes
⇒ Enclosed plant, free from odours.
⇒ Minimal excess sludge production.
⇒ Water recycle.
⇒ Small foot print.

The paper will give a general overview of different MBR techniques currently being used in Europe and also explore how the advantages of MBR can be retrofitted onto existing conventional activated sludge plants. The MBR process can be categorised into two main areas:

- The submerged technique where membranes are inserted into the BioReactor vessel. Generally used for high flow rates and resulting in low flux rates.

- Conventional cross flow UF technique with high flow velocities and flux rates. Very much suited to high COD strength waste streams.

Extensive UK trials in the Food industry have been undertaken using a dedicated industrial base pilot plant and results will also be presented in the paper.

The MBR process in Europe has been developed by Wehrle Werk AG, Emmendingen, Germany and they have 10 years of MBR experience, covering 40 installations. Stork Protech UK and Wehrle have combined forces and operate in the UK as a joint venture.

OPERATION OF A ZERO DISCHARGE WOOD PULP EFFLUENT TREATMENT PLANT

G.Bateman

Esmil Process Systems Ltd
Westfields
London Road
High Wycombe
HP11 1HA

Conventional processes for the treatment of effluent generated during the manufacture of medium density fibreboard (MDF) cannot comply with increasingly stringent environmental legislation. Advances in polymer chemistry combined with membrane technology have enabled the recovery of both water and soluble organic compounds. This has resulted in the realisation of a ZERO DISCHARGE process.

Benefits of the membrane system include: Contaminated effluent to pure water in one process stage, soluble product recovery – dissolved metals, organics and inorganics, robust process - not susceptible to thermal and toxic shocks

Following the success of the world's first commercialised system at an MDF mill in Wales, a further seven plants have been installed across Europe, making this process something of an industry standard. The operational experience gained from these plants has placed Esmil in a strong position to offer expertise in the field of water/product recovery and reuse.

Process Philosophy - Wood pulping effluent from the refining process is flocculated, then pumped to a filter press, which produces a filter cake with a dry solids content in excess of 50%. Filtrate passes through a multi-media filter. The filtered effluent is then fed into the RO membrane and finally an optional carbon filter. Concentrate from the RO membrane is stored in a dedicated holding tank, prior to recycling. All dirty backwash / cleaning waters from the filter press, multi-media filter and RO membrane system are recovered and returned to the head of the plant. All solid and liquid phase outputs are recoverable, thereby resulting in a zero discharge plant. The Esmil Plant is substantially automated and an operator can be fully trained within a matter of days.

The plant is largely mechanical with few process stages, reducing the capital investment requirement compared to other systems such as biological treatment.

This process is currently being considered as a case study under the DTI's Environmental Technology Best Practice Programme and has already this year won Esmil Process Systems the Queen's Award for Environmental Achievement.

INTEGRATION OF MAINTENANCE AND OPERATION INTO THE DESIGN OF REVERSE OSMOSIS MEMBRANE NETWORKS

H. J. See, V. S. Vassiliadis and D. I. Wilson

Department of Chemical Engineering,
University of Cambridge,
Pembroke Street,
New Museums Site,
Cambridge, CB2 3RA,
United Kingdom.

The selection of the configuration of a reverse osmosis (RO) membrane network is an important part of the design process, as the configuration dictates the flexibility of the system to variation in feed parameters and the loss of efficiency caused by fouling. Design methods have usually paid little attention to maintenance and control issues presented by membrane fouling and deterioration owing to the difficulty in determining the actual operating cost of a system subject to serious fouling and consequent cleaning.

A common practice to account for fouling consideration in the design process is to assume cleaning-in-place (CIP) operations based on pre-determined regeneration maintenance cycles. However, these cycles are empirically developed without exact understanding of effects of fouling, nor consider the implications of downtime and imperfect cleaning. This may result in higher capital costs if the fouling effect is overestimated, and higher operating costs due to frequent regeneration if the fouling effect is underestimated. The cleaning costs are strongly affected by the cleaning schedule used, so that complete optimisation of the design requires consideration of the impact of the network design on the cleaning schedule. Furthermore, replacement scheduled in advance has been acknowledged as the most effective way to control costs and maintain plant capacity consistency within design specs.

This paper builds on a recently developed method for determining the optimal cleaning schedule in a given RO network subject to fouling. The scheduling problem is formulated as a mixed-integer nonlinear programming (MINLP) problem that can include control action and membrane degradation. The problem is solved based on algebraic equations and time discretisation methodology under the general algebraic modelling system (GAMS) environment using the Outer-Approximation/Extended-Relaxation algorithm. Based on the algorithm, the objective function (which account for factors related to membrane regeneration and operating constraints) is minimised sequentially as a nonlinear programming (NLP) subproblem using MINOS and as a mixed-integer linear programming (MILP) master problem using CPLEX.

This flexible scheduling solver has been integrated into the larger design optimisation problem and the integrated problem solved using MINLP techniques. The size of the problem requires the use of approximate solutions to the design equations in the initial stage; these approximations are replaced by more exact forms in searching for the optimum configuration. The paper presents some initial results from this combined approach, using a previously published case study featuring a less flexible method. The sensitivities of the discrete time intervals and the fixed cost incurred on each cleaning period regardless of module quantities to be cleaned are also discussed.

MICROFILTRATION AND REVERSE OSMOSIS OF KNOSTROP FINAL EFFLUENT

M. Barton

Yorkshire Water Services,
Western Way,
Halifax Road,
Bradford,
BD6 2 LZ

Following the water shortages of recent summers, in particular 1995, Yorkshire Water wished to investigate potential new sources of water. Treating sewage effluent could potentially be lower cost than constructing a new raw water storage reservoir. A second reason for carrying out this work was to maintain competitive advantage and to offer Yorkshire Water's industrial customers a choice. An on-site treatment plant which can treat industrial effluent to a sufficiently good quality to enable it to be re-used as process water is an alternative to the provision of potable water and sewerage services. Treated sewage effluent could also potentially be used as on-site grey water, for polymer make-up or screenings washing.

The process chosen for this trial was continuous microfiltration (CMF) followed by reverse osmosis (RO).

The aim of the trial was to treat Knostrop STW effluent to potable water quality, establishing capital and operating costs and evaluating operational problems and risks associated with the process.

The programme of work involved maximising the flow through the pilot plant, optimising chemical usage, and challenging the plant with cryptosporidium and a selection of organic chemicals.

The CMF unit achieved a maximum throughput of 3.5 m^3/hr and the RO plant 2.5 m^3/hr. Several operational problems were experienced during the trial, including biofouling, iron fouling and problems with hypochlorite dosing. The quality of the effluent produced was excellent and has been described as "better than drinking water" and "poor quality distilled water". The cryptosporidium challenge achieved removal rates of greater than 4 log.

The trial of CMF and RO has proven that this technology is appropriate for the production of potable quality water from sewage effluent.

MODELLING TEMPERATURE AND CONCENTRATION POLARISATION IN ULTRAFILTRATION OF NON-NEWTONIAN FLUID UNDER NON-ISOTHERMAL CONDITIONS

Sergei P. AGASHICHEV

Department of Chemical Engineering ,
D. Mendeleev University of Chemical Technology,
Muisskaja Sq. 9,
Moscow 125047,
Russia

Development of new generation of membrane processes, in particular, in biotechnology and oil processing, requires development of advanced and more complex methods of calculation where phenomena of temperature and concentration polarization have to be described. Proposed method is based on hybridization of submodels for baro- and thermo-membrane operations. Submitted paper covers the following aspects: (A) temperature field; (B) concentration field; (C) non-Newtonian behaviour of fluid. The model is applicable for channels with plate-and-frame configuration. The proposed solution is based on the following assumptions: (1) flow is to be incompressible, laminar, continuous with uniform density field under the steady-state conditions; (2)fluid manifests non-Newtonian behaviour, viz.: apparent viscosity- $\mu(z)$ is a power function of shear rate The manuscript covers *pseudoplastic* and *dilatant* fluids; (3) fluid properties are independent of applied pressure and time under shear; (4) velocity profile being derived through integrating the momentum balance equation for non-Newtonian fluid has the following mathematical formulation $u(z) = U_{MAX}\left\{1 - (1 - z/H)^{(m+1)/m}\right\}$; (5)Temperature profile was approximated by parabolic function $\left(t(\varepsilon) - t_{1M}\right)/\left(t_1 - t_{1M}\right) = 1 - \varepsilon^2$, where ε-dimensionless coordinate with scale unit being equal to temperature layer thickness; The ratio of temperature (δ_T) to hydrodynamic (δ_W) boundary layer was estimated as $\delta_T/\delta_W = Pr^{-1/2}$; (6) Concentration profile was approximated by parabolic function $\left(c(\theta) - C_1\right)/\left(C_{1M} - C_1\right) = \theta^2$, where θ-dimensionless coordinate with scale unit being equal to concentration layer thickness. The ration of concentration to hydrodynamic boundary layer was estimated as $\delta_C/\delta_W = Sc^{-1/3}$; (7) No slip at the wall;

Equations for temperature and concentration boundary layers are the core of the model. Integration of these equations over control volume has been carried out. Proposed model permits the analysis of the influence of mass flow, bulk concentration, tangential velocity, transmembrane flux , membrane rejection and geometry parameter on configuration of concentration and temperature profile. The proposed solution can be built into complex algorithm of membrane process. The solution does not contain digital integration. The paper includes the developed algorithm, an array of calculated data and graphical visualization of solution.

NOVEL METHODS OF HOLLOW FIBRE MEMBRANE INTEGRITY MONITORING

Dr. S. Williams

Thames Water Research & Technology,
Spencer House,
Manor Farm Road,
Reading, RG2 0JW,
UK

Dr. A.J. Merry and C.V. Meadowcroft

Leopold-PCI Membranes,
Laverstoke Mill,
Whitchurch,
Hampshire, RG27 8NR,
UK

ABSTRACT

The recent widespread use of similar membranes for water treatment in Europe and North America has been largely driven by a need for improved disinfection, in particular the requirement to deal with Cryptosporidium. UltraBar is a novel hollow fibre membrane system, of which its main applications are in the single stage treatment of ground and surface waters to potable standards. As the system performs as a physical barrier, the most important factor is to ensure the integrity of the system. It is essential to confirm that the barrier provided by the membranes and components within the system is intact. Water treatment other than through an intact membrane system will allow the passage of potentially harmful organisms into supply. In the case of Cryptosporidium, only a very small number of organisms may be sufficient to cause infection. Such integrity breaches may take the form of membrane fibre breakages, pin hole imperfections, potting defects or system O-ring and seal defects.

In developing an integrated integrity monitoring system for the UltraBar process, a number of methods are currently being trialed at pilot scale. These include physical methods such as a water displacement test, a vacuum test and an air pressure test and a continuous online method using particle counting for an indication of permeate quality.

The water displacement test is based upon the theory that below a given applied pressure, the rate of air passage from the feed side to the permeate side of an intact membrane will be a very small and due mainly to diffusion. However, in the case of an integrity breach, the rate of air passage across the membrane is much higher. For the test, the feed side of the membrane elements are drained and the permeate side is left fully flooded with water. Air pressure to approximately 0.6 bar is applied to the feed side of the membrane, and the passage of air across the membrane results in the displacement of water from the permeate side. By measuring the rate at which this water is displaced, the integrity of the membrane is monitored. This method of integrity testing is currently in its developmental stages and is undergoing onsite pilot testing.

The initial results obtained on this test were generated using a single 40" XFlow element. The results were then applied to a full scale membrane stack consisting of 84 x 40" membrane elements. in order for the test to be sensitive enough to detect a single broken fibre within an entire membrane stack of 56 elements, it was calculated that the rate of water displacement due to air flow through a broken fibre on an compromised element would need to be approximately 100 times greater than the rate of water displacement due to air diffusion in each of the 83 intact elements. The results obtained on a single 40" element indicated that at an applied pressure of 0.6 bar, the measured R values ranged from 300 to 980. These results indicate that the sensitivity of the water displacement test is high enough to enable the detection of a single broken hollow fibre in over 800,000 intact fibres. Testwork carried out at a Thames Water site using a pilot plant containing 16 x 60" XFlow elements has produced comparable results.

The vacuum test is carried out only on elements and complete pressure vessel assemblies which have been removed from production and fully drained. The test is based upon similar theory to the water displacement test in as far that when a vacuum of 0.2 bar absolute is applied to the permeate side of a fully drained and intact membrane element, the rate of decay in the vacuum will be very slow and due only to the passage of air across the membrane from diffusion. Where an integrity breach is present however, any vacuum applied to the permeate side will rapidly decay due to the influx of air through a broken fibre from the open feed side of the membrane. The mean rate of vacuum decay measured ranged from 4.0 mBar per minute on an intact membrane element to 20 mBar per minute on an element with one broken fibre.

Air pressure is also used in the location and repair of individual broken fibres. This test takes place on individual membrane elements which have been removed from their pressure vessels and identified as having a breach of integrity. The membrane is fully immersed into a tank of water and flushed with water in order to remove any air from its feed side. Compressed air is applied to the permeate side of the submerged membrane. If a broken fibre is present, this will be indicated by a stream of air bubbles escaping from the broken fibre at both ends of the element. The element is repaired by the insertion of plastic repair pins into both ends of the broken fibre. This acts by sealing the fibre, therefore preventing any further flow of raw water into the fibre. The membrane element can then be placed back into process and reused.

A novel multiple headed particle counting system has also been developed and is currently undergoing testing at pilot scale. The system monitors for particle breakthrough as a method of monitoring membrane integrity and allows for a single particle counting head to be fitted to the permeate offtake of each pressure vessel. The positioning of a single counting head per vessel allows for immediate identification of within which pressure vessel an integrity breach has occurred. This vessel can then be isolated from process for repair. Each set of particle counting heads is connected to a common control panel which houses the collective electronics and alarms.

COMPARISON OF CHEMICAL PRETREATMENT METHODS FOR NANOFILTRATION OF COLD, SOFT AND HUMIC WATERS

J. Yli-Kuivila, R. Liikanen and R. Laukkanen

Laboratory of Environmental Engineering
P.O. Box 5300
FIN–02015 HUT
Finland

The applicability of nanofiltration (NF) to drinking water production was studied in seven Finnish surface waterworks with a pilot filter (1 m^3/h) using four spiral-wound 40-40 elements and circulation. The organic quality of NF permeate was excellent in all the cases: TOC was consistently below the detection limit (0.3 mg/l), assimilable organic carbon very low, and disinfection by-product formation potential reductions high.

Studies concerning fouling, pretreatment and maintenance practices have been continued in the Espoo waterworks, where the raw water quality is typical for the area: cold (0.1–23 °C), soft (<1 °dH) and humic ($KMnO_4$ 25–60 mg/l). The variation of quality is considerable. The use of polyaluminium chloride (PACl) and two ferric salts was optimised according to the MFI and SDI membrane fouling indices. Sedimentation (70 l batch process) and dissolved air flotation (DAF, 6.5 m^3/h, 8 m/h) with rapid sand filtration were studied with cold raw water (<2 °C). Sedimentation was replaced by DAF in the Espoo waterworks and NF was tested with both pretreatment methods.

Figure1 illustrates the results of MFI optimisation, which is equivalent to minimisation of SDI. According to recent MFI values, the conventionally optimised full-scale process works better than the pilot-scale unit. The following conclusions can be drawn from the results of the study:

- Using conventional coagulant dosages (170 $mmol_{Me3+}/m^3$, flocculation pH Al: 5.7, Fe: 4.5) better results were achieved with PACl compared to ferric salts
- Better results were obtained at higher ferric salt dosages (300 $mmol_{Fe}/m^3$)
- The NF-pilot unit performed slightly better with DAF than with sedimentation

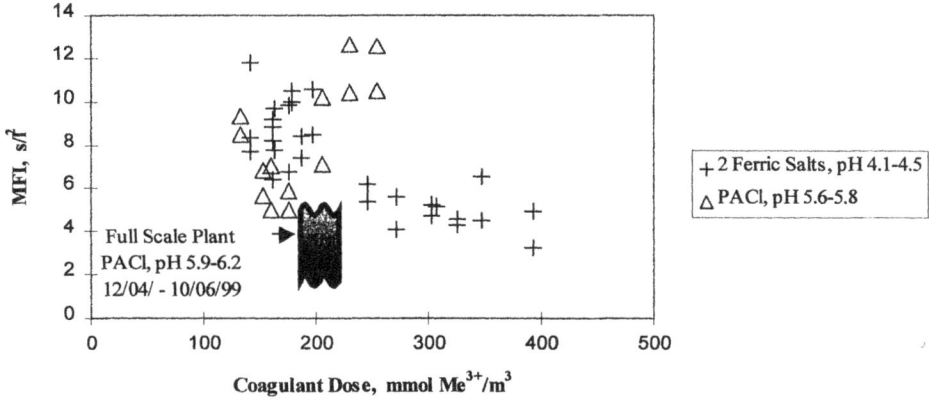

Figure 1 *Optimised MFI in flotation – sand filtration pilot studies 04/01/ - 31/03/99*

IN-SITU ULTRASONIC MEASUREMENT OF FOULING AND CLEANING PROCESSES IN SPIRAL-WOUND MEMBRANE MODULES

G.-Y. Chai, A. R. Greenberg and W. B. Krantz

Departments of Chemical and Mechanical Engineering
NSF I/U CRC for Membrane Applied Science and Technology
University of Colorado
Boulder, CO 80309-0424

1 INTRODUCTION

Fouling is a major problem in membrane processes that significantly limits the further utilization of membranes in water treatment and other liquid separations. The lack of suitable methods to monitor fouling and cleaning under realistic operating conditions has hindered the development of strategies to improve membrane and membrane module resistance to fouling. While the occurrence of fouling is usually identified by a marked decrease in permeate flux and quality, this decline is not always attributable to fouling. In addition, measurements of permeate flux cannot be obtained when the membrane module is undergoing chemical cleaning. With these limitations in mind, the overall objective of this study is the adaptation of ultrasonic time-domain reflectometry (UTDR) for the real-time, *in situ* measurement of $CaSO_4$ fouling and subsequent chemical cleaning in commercial RO membrane modules under a range of operating conditions.

2 EXPERIMENT

The experimental flow system consisted of a commercial spiral-wound RO membrane module, a temperature-controlled feed tank, and standard devices for the control and measurement of pressure, feed concentration, flow rate and temperature. Membrane flux and rejection were measured at regular intervals using standard techniques. The ultrasonic hardware included high-frequency transducers, a pulser-receiver and a digital oscilloscope. The transducers were mounted on the external surface of the module, and the sampling rate for the oscilloscope was usually set at 20 nsec. The range of operating conditions evaluated included 0-2.72 MPa for the upstream pressure, 10-30 °C for temperature, 0.3-2.0 l/min for flow rate and 0-1.5 g/l for $CaSO_4$ feed concentration.

3 RESULTS

Representative results that show the systematic relationship between the permeate flux and UTDR signal amplitude are presented in Figure 1 for the pretreatment conditioning,

CaSO$_4$·2H$_2$0 fouling, and cleaning stages of operation at an upstream pressure of 0.68 MPa, and a constant temperature of 20 °C. In the fouling phase the relative permeation flux decreased by 28% while the relative signal amplitude increased by about the same magnitude. During the subsequent cleaning phase, the flux increased such that, by the end of the cleaning cycle, the flux was statistically indistinguishable from its initial value. Interestingly, the UTDR signal amplitude decreased over this same time period. Indeed, by the end of the cleaning cycle the UTDR signal amplitude appeared to be approaching the initial level measured prior to the onset of fouling. Also note that the UTDR signal amplitude values could be obtained continuously during the cleaning cycle, whereas this was not possible for the permeate-flux measurements. In order to determine the applicability of the UTDR technique for monitoring a commercial spiral-wound module under realistic conditions, the effect of changes in the operating parameters on the signal amplitude response were studied. These results suggest that only the operating pressure and temperature significantly affect the signal amplitude.

4 CONCLUSIONS

Overall, the results of this study indicate that UTDR shows significant potential for monitoring fouling in spiral-wound membrane modules. The data also indicate the importance of establishing a suitable calibration protocol so that accurate measurements can be obtained despite normal variations in the membrane module operating parameters.

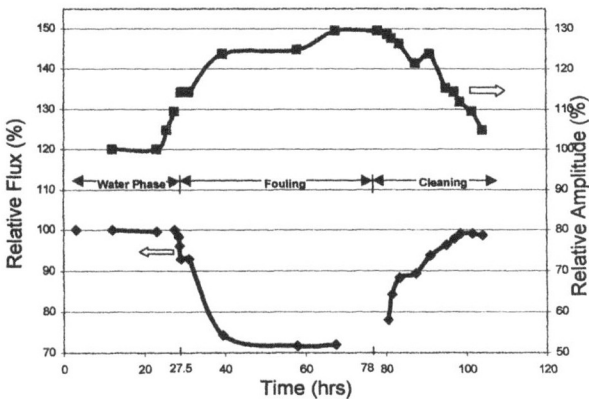

Figure 1. *Representative results during module operation showing correspondence between permeate flux and UTDR signal amplitude measurements during the pretreatment conditioning, CaSO$_4$·2H$_2$0 fouling, and cleaning phases.*

A NOVEL WAY TO TREAT TEXTILE WASTEWATER WITH NANOFILTRATION AND ADSORPTION

Prof. Dr.-Ing. T. Melin and Dipl.-Ing. L. Eilers

Institut für Verfahrenstechnik der RWTH Aachen (IVT),
Turmstr. 46,
52056 Aachen,
Germany

Wastewater from textile dying operations has to be treated not only because of its colour, but also because of its high content of poorly biodegradable organics. A treatment process therefore has to provide both a low cost sink for the organic load and a near-complete removal of colour and critical components. Single step processes can achive only one of these goals, PAC (powdered activated carbon) has to be used in high (i.e. expensiv) quantities for complete complete colour and DOC-removal whereas NF (nanofiltration) can achieve sufficient colour removal only at the cost of a large volume of concentrate, which cannot be further reduced due to limited colour retention and / or excessive osmotic pressure. By performing adsorption on inexpensive powdered adsorbents (e.g. lignite coke dust) in an NF-module we developed a new hybrid process, which avoids the weaknesses of the single step processes while fully using their strengths: NF as a colour, DOC and solids barrier and the lignite coke dust (adsorbent) as a cost effective sionk for the organic poad.

Basically, Nanofiltration presents a promising alternative for rejecting colour - but a highly concentrated brine is inevitable. For reducing the concentrates high recovery rates and, consequently, high colour concentrations are mandatory. But high concentrations lead to reduced driving forces in the membrane process - low permeate fluxes need to be compensated with high membrane areas. Furthermore, high concentrations result (constant membrane selectivities assumed) in higher permeate colour concentrations. As a result, the reduction of the concentrates is limited because of economical and ecological reasons.

The process combination Nanofiltration and adsorption at powdered adsorbents (for example lignite coke dust) in the membrane module seems to solve the dilemma of the standalone processes Nanofiltration and adsorption by combining the advantages of the single processes. In a project funded by the "Bundesministeriums für Wirtschaft" via the "Arbeitsgemeinschaft industrielle Forschungsvereinigung "Otto von Guericke" e.V. (AiF-Vorhaben-Nr. 112 42)" this process combination was developed.

The lignite coke dust not only reduces the adsorbable components (i.e. colour) in the wastewater but also improves the performance of the Nanofiltration. In detail, the adsorbent performs as

■   a sink for adsorbable components (i.e. colour)

■   a filter-aid: the adsorbent supports the build-up of less compact fouling layers on the membrane with little hydraulic resistance

■   a pre-coat: the adsorbent layer directly on the membrane supports an easy removal of the fouling layer for example by means of gas sparging.

The fouling layer consists of the adsorbent itself and the rejected components of the wastewater (i.e. colour) and can be removed from the membrane periodically (by means of gas sparging) thanks to the pre-coat function of the adsorbent. This periodic module flushing can be applied if membrane modules with open channels are used. The achieved improvement in membrane performance allows a drastic reduction of the cross-flow velocity and, therefore, energy consumption. It is crucial that the effectiveness of the periodic membrane cleaning is not limited by the concentration of the adsorbent - concentrations of up to 10 g/l were tested.

www.ingramcontent.com/pod-product-compliance
Lightning Source LLC
Chambersburg PA
CBHW021429180326
41458CB00001B/195